Lecture Notes on Data Engineering and Communications Technologies

Volume 256

AF803859

Series Editor

Fatos Xhafa, Technical University of Catalonia, Barcelona, Spain

The aim of the book series is to present cutting edge engineering approaches to data technologies and communications. It will publish latest advances on the engineering task of building and deploying distributed, scalable and reliable data infrastructures and communication systems.

The series will have a prominent applied focus on data technologies and communications with aim to promote the bridging from fundamental research on data science and networking to data engineering and communications that lead to industry products, business knowledge and standardisation.

Indexed by SCOPUS, INSPEC, EI Compendex.

All books published in the series are submitted for consideration in Web of Science.

Smain Femmam · Pascal Lorenz
Editors

Recent Advances in Communication Networks and Embedded Systems

Proceedings of 8th International Conference on Communication and Network Technology

Springer

Editors
Smain Femmam
University of Haute Alsace
Mulhouse, France

Pascal Lorenz
University of Haute Alsace
Colmar, France

ISSN 2367-4512 ISSN 2367-4520 (electronic)
Lecture Notes on Data Engineering and Communications Technologies
ISBN 978-3-031-92448-4 ISBN 978-3-031-92449-1 (eBook)
https://doi.org/10.1007/978-3-031-92449-1

© The Editor(s) (if applicable) and The Author(s), under exclusive license to Springer Nature Switzerland AG 2025

This work is subject to copyright. All rights are solely and exclusively licensed by the Publisher, whether the whole or part of the material is concerned, specifically the rights of translation, reprinting, reuse of illustrations, recitation, broadcasting, reproduction on microfilms or in any other physical way, and transmission or information storage and retrieval, electronic adaptation, computer software, or by similar or dissimilar methodology now known or hereafter developed.
The use of general descriptive names, registered names, trademarks, service marks, etc. in this publication does not imply, even in the absence of a specific statement, that such names are exempt from the relevant protective laws and regulations and therefore free for general use.
The publisher, the authors and the editors are safe to assume that the advice and information in this book are believed to be true and accurate at the date of publication. Neither the publisher nor the authors or the editors give a warranty, expressed or implied, with respect to the material contained herein or for any errors or omissions that may have been made. The publisher remains neutral with regard to jurisdictional claims in published maps and institutional affiliations.

This Springer imprint is published by the registered company Springer Nature Switzerland AG
The registered company address is: Gewerbestrasse 11, 6330 Cham, Switzerland

If disposing of this product, please recycle the paper.

Conference Committee

Advisory Committee

Ljiljana Trajkovic, Simon Fraser University, Canada

Conference Chair

Mahmoud Shafik, University of Derby, UK

Conference Co-chairs

Alexey Vinel, Karlsruhe Institute of Technology (KIT), Germany
Maurici Ruiz, University of the Balearic Islands, Spain

Program Chairs

Gang-Len Chang, University of Maryland, USA
José Paulo Lousado, Polytechnic Institute of Viseu, Portugal

Publicity Chairs

Angel-Antonio San-Blas, Miguel Hernandez University of Elche, Spain
Thomas Hanne, University of Applied Sciences and Arts Northwestern Switzerland, Switzerland
Yasmany García-Ramírez, UTPL, Ecuador

Technical Program Committees

Stavros Shiaeles, University of Portsmouth, UK
Gururaj H. L., Manipal Institute of Technology, India
Pascal Lorenz, University of Haute Alsace, France
Mohamed Firdhous, University of Moratuwa, Sri Lanka
Luca Davoli, University of Parma, Italy
Daniele Codetta Raiteri, Università del Piemonte Orientale, Italy
Smain Femmam, Haute-Alsace University, France
Apostolos Xenakis, University of Thessaly, Greece
Napa Sae-Bae, SWU, Thailand
Rinku Basak, American International University-Bangladesh, Bangladesh
Khondker Shajadul Hasan, University of Houston, Clear Lake, USA
Banphatree Khomkham, SWU, Thailand
Dimitris Kanellopoulos, University of Patras, Greece
Apostolos Gkamas, University of Ioannina, Greece
Gowtham M., National Institute of Engineering, Mysuru, India
Thien Wan Au, Universiti Teknologi Brunei, Brunei, Darussalam
Nasimuddin, Institute for Infocomm Research, Singapore
Xiaoxuan Wang, Beijing Jiaotong University, China
Jiajia Jiang, Tianjin University, China
Kwanho You, Sungkyunkwan University, Korea
Kasturi Vasudevan, IIT-Kanpur, India
Qamar Nawaz, UAF, Pakistan
Haibo Yang, Shenyang University of Technology, China
Chuan-Ming Liu, National Taipei University of Technology, Taiwan
Abdul Halim Zaim, Istanbul Technical University, Türkiye
Djalal Benyahya, Ecole Militaire Polytechnique, Algeria
Lazoul Mohamed, Ecole Militaire Polytechnique, Algeria

Preface

The International Conference on Communication and Network Technology (ICCNT) is a major event for academics, researchers, and industrialists who are engaged in Communication and Network Technology research. Held annually since the late 2010, the conference is renowned as a friendly and inclusive platform that brings together a broad community of researchers who share a common goal: developing and managing engineering and technologies revolution of Communication and Network Systems, and operations which are key to sustaining the success of Communication and Network Technology industries. For more than a decade, ICCNT has been the main Communication and Network research conference organized overall across EU and the world, successfully, bringing researchers, academics, and industrialists together to share their knowledge, know-how and experiences. Initiated as a Communication and Network National workshop by ICCNT Consortium of well-known Universities key professors in Communication and Network applied research, and it became an International Conference in 2016.

ICCNT Consortium is an independent body established in 2010 and led by Prof. Mahmoud Shafik. Its main aim is to promote Communication and Network engineering and technology education, training, research and knowledge transfer, and exchange. To achieve this, the conference chairmen, consortium, and committee members maintains a close association with international bodies concerned with the training and continuing development of professional engineers and technologists, while responding to the appropriate consultative and discussion of documents and other initiatives. The conference also plans major industrial visits to, Small to Medium Enterprises (SME's) and Large Enterprise (LE's) industries to enhance conference participants knowledge and experience of the existing and future industrial revolution (i.e., 5G, and beyond, digital transformation, industry 4.0 and beyond, autonomous technology, 2D and 3D data streaming…etc.) and its impact on Communication and Network Technology market.

In 2016 the ICCNT conference was given the title International (ICCNT) to reflect the current trends in Communication and Network engineering and technology and to promote the exchange of research, engineering, and technology application experiences internationally. The ICCNT has been taken place in the previous years at the following countries:

ICCNT 2016 China
ICCNT 2017 France
ICCNT 2018 Sweden
ICCNT 2019 Netherlands
ICCNT 2020 VR Online
ICCNT 2021 VR Online
ICCNT 2022 France
ICCNT 2023 Spain
ICCNT 2024 Italy

Prof. Mahmoud Shafik
Conference Chair
University of Derby
Derby, UK

Contents

Emerging Network Technology and Information Security

Software-Based Information Communication and Application Technology

Conception and Simulation of Photoconductive Terahertz Emitters for Wireless Communication

H. Zeraoula, D. Benyahia, M. Mbarek, and M. Lazoul

Abstract Terahertz (THz) technology has emerged as a potential candidate for 5G and 6G communication. This is due to its prominent characteristics, such as ultra-high data transfer rates and large bandwidth. However, there are many drawbacks that impede the use of this technology, especially the conception and the design of THz sources. Photoconductive antennas (PCAs) are considered as the best candidate that allows the conception of ultrafast wireless communication systems. In the present work, we investigate and analyze two theoretical models: analytical and that based on equivalent circuit, which are used to characterize the generated photocurrent and the radiated THz pulse in the PCA. Besides, we discuss each model principle to identify the most convenient situation for applying it efficiently and correctly. Furthermore, we investigate the PCA parameters effect on the THz radiation for both: the photoconductive material and the exciting femtosecond laser. Simulation results allow the well perception of the physical interpretation for the radiation mechanism in each model, which leads to realize the consequence of misapplying the convenient model for the design, which is still the case of recent works. Moreover, the concepted PCA allows to have a wide bandwidth, that facilitates the realization of future wireless communication systems.

Keywords Terahertz · Photoconductive antenna · Equivalent electric circuit model · Wireless communication

1 Introduction

Wireless communication has become an indispensable necessity in the world due to its numerous applications in different domains, such as: security, public, and economy [1]. However, this communication type needs a large bandwidth, to ensure an ultrafast network. This feature can be guaranteed by the terahertz technology.

H. Zeraoula · D. Benyahia (✉) · M. Mbarek · M. Lazoul
Laboratoire Systèmes Lasers, Ecole Militaire Polytechnique, Algiers, Algeria
e-mail: djalal.benyahia@emp.mdn.dz

© The Author(s), under exclusive license to Springer Nature Switzerland AG 2025
S. Femmam and P. Lorenz (eds.), *Recent Advances in Communication Networks and Embedded Systems*, Lecture Notes on Data Engineering and Communications Technologies 256, https://doi.org/10.1007/978-3-031-92449-1_1

The THz band is located between the microwave and the infrared regions, in other words, in the frequency range of 0.1 to 10 THz [2]. However, this band has not been well investigated due to the absence of adequate sources and detectors [3, 4]. Nonetheless, THz technology has lately known a remarkable attention thanks to its unique properties, namely: high resolution, better penetration, and non-ionizing property [5]. Hence, THz finds several applications in security and defense [6], medical imaging [7] and bio-chemical spectroscopy [8]. Besides, THz technology has an important role in wireless communication which is appropriate for 6G and beyond due to its large bandwidth (1–10 THz), and ultra-fast data transfer rate (1–10 Tbps) [9].

The generation of THz waves is a challenging task. There are numerous types of THz sources like quantum cascade lasers, IMPATT diode, spintronic materials and Schottky diode multiplier, and so forth [10]. However, these latter are generally expensive, bulky and they need laboratory-like environment for real time experiments. One solution of these drawbacks is the photoconductive antennas. These sources have various advantages, which contribute to their reputation. For example, they are microscale, works at room temperature, and can be utilized as an emitter or a detector [11, 12]. Besides, unlike typical radio frequency (RF) antennas, the photoconductive ones need the use of more than one software to simulate their whole functioning, due to the different physical mechanisms that are involved in this antenna type [13].

In this work, we aim to study two important models used to characterize PCAs, which concerns analytical and equivalent electric circuit models. We describe and explain each model principle to apply it conveniently, and then to evoke the concern of model misemploying. Indeed, some research groups [14] have applied analytical model based on Maxwell equations for small gap PCA, while the used model is tailored for the large gap PCA. Moreover, we simulate the effect of photoconductor material and the influence of the used femtosecond laser as exciting source on the performance of the generated photocurrent and consequently the radiated THz pulse to optimize the design of the PCA for wireless communication systems. In addition, we simulate various situations for the two models and discuss the obtained results concerning the photocurrent, radiated THz pulse and the obtained THz spectrum.

2 Photoconductive Antenna for THz Radiation Principle

In THz systems, PCA serves as both an emitter and a detector. For the first one, it transforms optical radiation into THz waves. On the other hand, it converts THz energy to electric energy when it acts as a detector. However, PCA analysis, simulation and measurement should be treated by different techniques due to their unique excitation mechanism. Auston switch [15] was the first PCA which was utilized to generate a pico-second pulse. This switch involves a gap in between two biased metallic electrodes, which are placed on the top surface of an ultra-fast photoconductor material, that acts as a Hertzian dipole. In addition, the first to report on

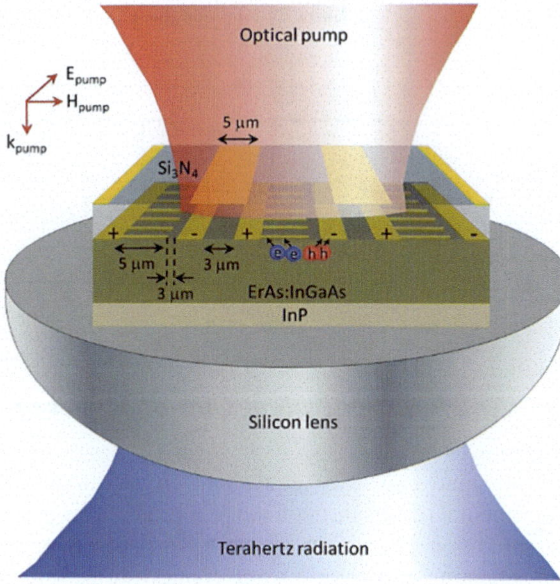

Fig. 1 Schematic diagram of a photoconductive antenna [5]

photoconductive antennas deposited on a LT-GaAs (low temperature grown GaAs) substrate were Warren et al. [16].

Figure 1 presents a schematic diagram and the working principle of a photoconductive antenna for a THz pulse generation. As can be seen, a DC biased metal dipole is deposited on an InP (Indium Phosphide) substrate. The exciting optical pulse, generated by a femtosecond laser, illuminates the antenna gap. Consequently, photocarriers (electron and holes) are generated into the photoconductive substrate. Later, a transient photocurrent is induced by accelerating the generated photocarriers in the presence of a DC bias electric field. This photocurrent drives the photoconductive antenna and then a THz radiation is emitted through the silicon lens [5]. It should be noted that a Si_3N_4 antireflection coating layer is utilized in order to improve the optical transmission through the substrate.

3 Theoretical Modeling of Photoconductive

Due to the variations in antenna gap size, different theories for the generation of THz photocurrent may exist to explain the radiation mechanism and match the obtained experimental results.

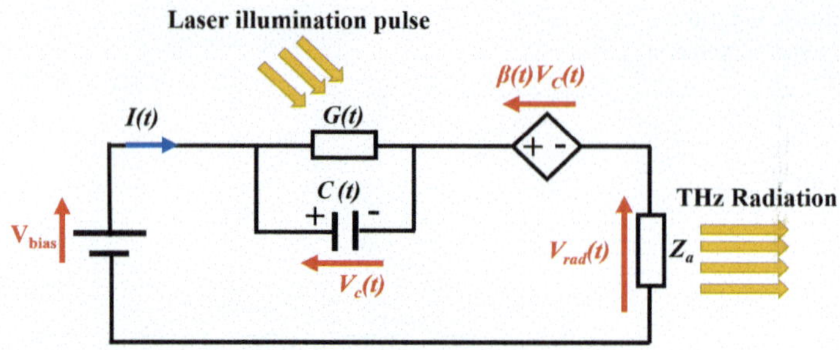

Fig. 2 Equivalent circuit model for PCA

3.1 Antenna Equivalent Circuit Model of THz PCA

For the small gap PCA, equivalent circuit model can describe well the obtained experimental results through the proposed model parameters, which explains mainly the saturation phenomenon due to the screening effect induced by both: radiation screening and space charge (polarization) screening [17]. Figure 2 shows the equivalent circuit model elements.

3.2 Analytical Model for Large PCA

Large-aperture PCAs have been used for generating high-power THz pulses. The main mechanism of THz generation in a large-aperture antenna is comparable to that in a dipole antenna. The generated photocurrent can be assumed as a surface current confined in a thin layer in the photoconductive gap [18]. Hence, the radiated THz field can be derived by using Maxwell equations:

$$E_{THz}(r,t) = -\frac{\mu S}{4\pi r}\frac{\partial J_s(t)}{\partial t} \quad (1)$$

where $J_s(t)$ is the surface (denoted S) current density given by:

$$J_s(t) = -\frac{\sigma_s(t) * E_{bias}(t)}{\frac{\sigma_s(t)*\eta_0}{1+\sqrt{\varepsilon_r}} + 1} \quad (2)$$

with $\sigma_s\ (t)$ is the surface conductivity of the photoconductive substrate, and E_{bias} is the electric field of the voltage bias.

4 Simulation Setup and Results Discussion

This section is divided into three parts. In the first one, we simulate the effect of various parameters on the small gap PCA performance, by using equivalent circuit model, while in the second part, the simulation of the large gap PCA based on the analytical model are presented, then we discuss the obtained results in the last part.

4.1 Equivalent Circuit Model for Small Gap PCA Simulations

The simulation in this part concerns the effect of: the applied bias voltage, the femtosecond laser power, and the antenna impedance. Table 1 illustrates the parameters values used in different cases.

Figure 3 shows the resulting voltages in the equivalent circuit for different values of the applied voltage bias. V_{gap} is the voltage across the gap sides represented by $V_c(t)$ in Fig. 1, V_{ant} is the voltage applied on the antenna equivalent impedance and V_{sc} represents the space charge screening effect voltage induced by the polarization resulting for the carriers separation. We can see clearly that the antenna voltage increases as the voltage bias goes up, which is due to the fact that by increasing the voltage bias, the acceleration of the carriers becomes greater. In addition, the screening effect voltage increases slightly, but it will be saturated after certain period for both voltage bias values.

Figure 4 illustrates the variation in circuit voltages for different values of the exciting laser power. It can be seen that the antenna voltage increases as the laser power goes higher. This can be explained by the generation of higher density of electron–hole pairs by the excitation through a higher laser power.

Figure 5a, b presents the circuit voltages variation (the optical to THz radiation efficiency) for different values of antenna impedance, respectively. One can notice

Table 1 Simulation parameters for small gap PCA

Parameter	Value	
	Fixed case	Variable cases
Applied bias (V)	30.00	10, 30 and 50
Mean laser power (mW)	30.00	20, 70 and 150
Antenna impedance (Ω)	65.00	30, 65 and 120
Pulse duration (fs)	100.00	/
Optical absorption coefficient (cm^{-1})	6000.00	/
Gap size (μm)	10.00	/
Relative permittivity	12.95	/
Carrier life time (ps)	1.00	/

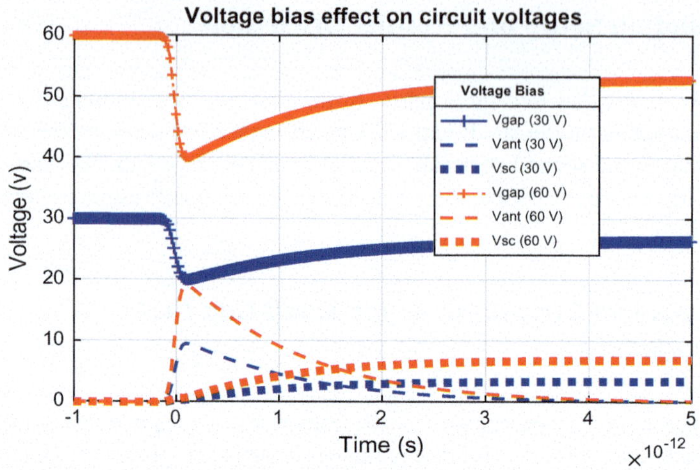

Fig. 3 Applied bias voltage effect on different circuit voltages

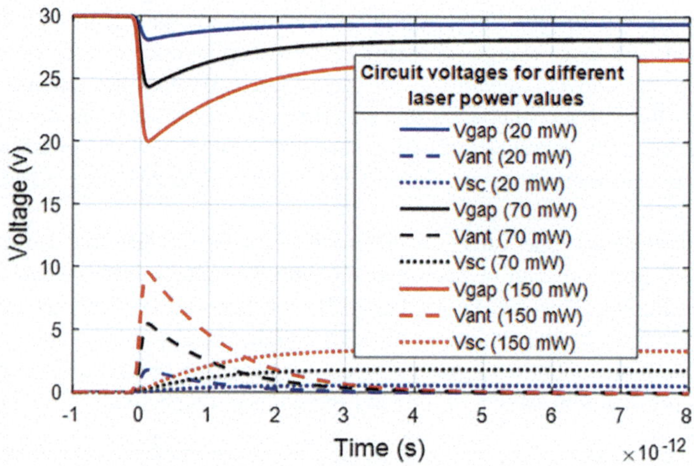

Fig. 4 Laser power effect on different circuit voltages

that the antenna voltage increases as the impedance of the antenna increases. On the other hand, the conversion efficiency is proportional to the antenna impedance value in the rise of the graph, then it will be inversely proportional as the average optical power increases beyond the saturation value.

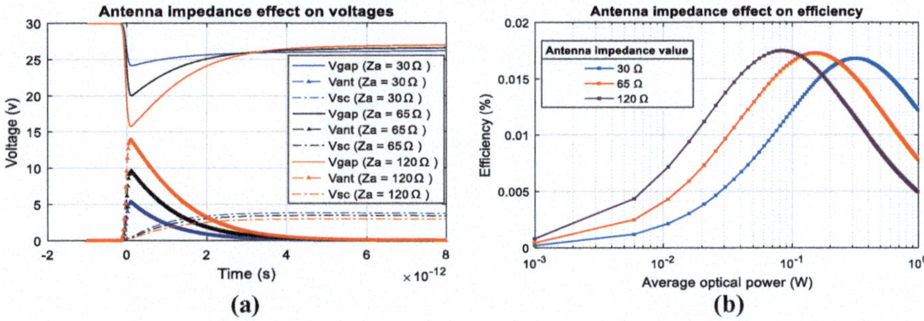

Fig. 5 Antenna impedance effect on **a** different circuit voltages and **b** optical to THz efficiency

4.2 Analytical Model for Large Gap PCA Simulations

The simulation in this part concerns the effect of: the applied bias voltage, the laser power, the laser pulse width and the photoconductor carrier lifetime. Table 2 illustrates the parameters values used in different cases.

The influence of the applied bias, the femtosecond laser power, and the laser pulse width, on the current density is analyzed through simulation results depicted in Fig. 6a–c, respectively. As can be seen, the current density increases as the applied bias increases, due to the high acceleration of the photogenerated carriers between the metallic dipoles. The same observation for the dependence of the photocurrent density versus the laser mean power, which can be explained by the fact that more photogenerated carriers are created in the photoconductive substrate. For the laser pulse width, it is inversely proportional to the current density amplitude, which is due to the fast recombination rate of the photogenerated carriers by Shockley–Read–Hall (SRH) mechanism.

Figure 7 shows the effect of the laser pulse width and the carrier lifetime on the THz spectrum of the radiated pulse. We can see clearly that the spectrum bandwidth

Table 2 Simulation parameters for large gap PCA

Parameter	Value	
	Fixed case	Variable cases
Carrier life time (ps)	0.30	0.1, 0.3, 0.8, 1.5
Pulse duration (fs)	100.00	40, 80, 120, 200
Applied bias (V)	60.00	10, 40, 80, 150
Mean laser power (mW)	30.00	10, 30, 50, 100
Optical absorption coeffecient (cm^{-1})	600000	
Gap size (μm)	300.00	
Relative permittivity	12.95	
Repetition frequency (MHz)	80.00	

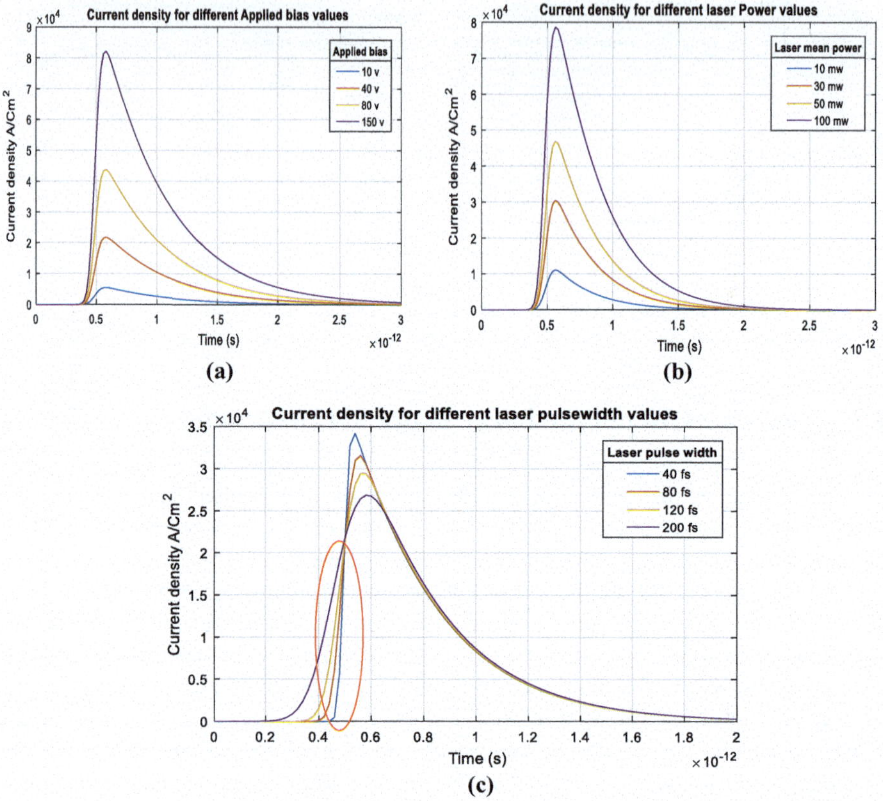

Fig. 6 The variation of the current density versus **a** the applied bias, **b** the laser power, and **c** the laser pulse width

increases as the laser pulse width decreases, while it decreases as the carrier lifetime increases.

4.3 *Results Analysis and Discussion*

From the radiation principle specificity for each PCA based on the gap size and the obtained simulation results, we can derive the following outcomes:

- The effect of some parameters is similar for the two antenna types, like the laser pulse width effect. In this resumed paper, we present simulation only for large gap (Fig. 7). It should be mentioned that the small gap simulation results are similar to that of large gap PCA. One can notice that the increase in laser pulse width reduces considerably the spectrum band width for the radiated THz pulse. Besides,

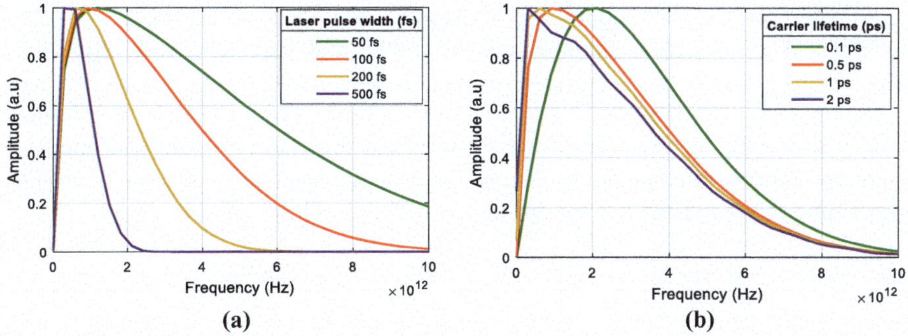

Fig. 7 The variation of the THz spectrum versus **a** the laser pulse width, and **b** the carrier lifetime

the carrier lifetime has a negligeable effect on the THz pulse bandwidth. Consequently, the THz pulse gets wider and shows noticeable asymmetry that reduces the frequency bandwidth, which is not good for the wireless communication.

- The radiation mechanism in small gap PCA is characterized by the screening effect which affects greatly the radiated pulse causing saturation phenomenon, the equivalent circuit model takes into consideration both: the radiation screening through the equivalent antenna impedance, and the space charge screening using the voltage source controlled by the equivalent capacity of the gap, thus as shown in Figs. 3, 4 and 5. The voltages in the equivalent circuit are changing in terms of the applied bias voltage, laser power and antenna impedance in different ways depending on the operation of each element in the circuit, which explains the reason why equivalent circuit is a good model for small gap PCA. On the other hand, for the large gap PCA, the screening effect, which was produced by the space charge, is relatively low because the crossed distance by the photogenerated carriers is too small comparatively to the gap size, that is why we notice from Fig. 6 that the current density increases linearly with the applied bias voltage and laser power. Furthermore, for the analytical model, the THz radiation is independent of the antenna equivalent impedance, because the analytical model based on Maxwell equations uses only the near field for current density estimation, so the analytical model is adapted strictly to the large gap PCA.

5 Conclusion

To conclude, PCA performances have been studied and simulated in order to define the optimized parameters for the conception of a performant wireless communication system. Therefore, two theoretical models (analytical model based on Maxwell equations and equivalent circuit model) that handle the generated photocurrent and the radiated THz pulse are studied and analyzed to be applied conveniently for PCA. The obtained results permit firstly, to explain the reasons for which the analytical model based on Maxwell equations is convenient only for large gap PCA, while the model based on equivalent circuit is adapted to the small gap PCA. Secondly, simulation results allow to assess PCAs performances concerning the exciting source and the photoconductive substrate for various parameters, to find the best compromise between the parameters with respect to the imposed constraints. Finally, the obtained results for the generated photocurrent and the THz spectrum provide more flexibility for defining the requested bandwidth for 6G or 5G wireless communication.

Acknowledgements This work is supported as part of the NATO-SPS Programme Multi-Year Project: N# G5396.

References

1. Rathinaasamy V, Thipparaju RR, Edwin NFB, Mondal S (2021) Interdigitated photocnductive terahertz antenna for future wireless communications. Microw Opt Technol Lett 64(12):1–8
2. Zeraoula H, Benyahia D, Lazoul M (2024) Proposing a refined and simplified equivalent circuit model for precise THz photoconductive antenna analysis. Opt Quant Electron 56:1055
3. Ferguson B, Zhang X (2002) Materials for terahertz science and technology. Nat Mater 1:26–33
4. Zeraoula H, Benyahia D, Lazoul M, Slimani C, Hadji Y (2024) Design and performance analysis of terahertz photoconductive antenna. In: 2024 International telecommunications conference (ITC-Egypt), pp 152–157. IEEE, Cairo, Egypt
5. Yardimici NT, Jarrahi M (2016) High-power, broadband terahertz radiation from large area plasmonic photoconductive emitters operating at telecommunication optical wavelengths. In: 2016 IEEE MTT-S international microwave symposium (IMS), pp 1–4. IEEE, San Francisco, CA, USA
6. Mangeney J (2012) THz photoconductive antennas made from ionbombarded semiconductors. J Infrared Millim Terahertz Waves 33(4):455–473
7. Bashirpour M, Forouzmehr M, Hosseininejad SE, Kolahdouz M, Neshat M (2019) Improvement of terahertz photoconductive antenna using optical antenna array of ZnO nanorods. Sci Rep 9(1):1–8
8. Rahmati E, Ahmadi-Boroujeni M (2018) Improving the efficiency and directivity of THz photoconductive antennas by using a defective photonic crystal substrate. Opt Commun 412:74–79
9. Rappaport TS, Xing Y, Kanhere O et al (2019) Wireless communications and applications above 100 GHz: opportunities and challenges for 6G and beyond. IEEE Access 7:78729–78757
10. Burford NM, El-Shenawee MO (2017) Review of terahtertz photoconductive antenna technology. Opt Eng 56(1):010901
11. Khiabani N (2013) Modelling, design and characterisation of terahertz photoconductive antenna. PhD thesis, Department of Electrical and Electronics Engineering, University of Liverpool
12. Lee Y (2008) Principles of terahertz science and technology, 1st edn. Springer, New York
13. Sajak A (2018) Numerical simulations of THz photoconductive antenna. PhD thesis, Department of Electrical and Electronics Engineering, University of Liverpool
14. Zhi-Jin Y, Wei S (2021) Radiation characteristics of terahertz GaAs photoconductive antenna arrays. Acta Phys Sin 70(24):248704
15. Auston DH, Cheung KP, Smith PR (1984) Picosecond photoconducting Hertzian dipoles. Appl Phys Lett 45(3):284–286
16. Warren AC, Katzenellenbogen N, Grischkowsky D, Woodall JM (1991) Subpicosecond, freely propagating electromagnetic pulse generation and detection using GaAs: as epilayers. Appl Phys Lett 58(14):1512–1514

17. Loata GC (2007) Investigation of low-temperature-grown GaAs photoconductive antenna for continuous-wave and pulsed terahertz generation. Ph.D. Thesis, Goethe-University, Frankfurt
18. Duvillaret L, Garet F, Roux J-F, Coutaz J-L (2001) Analytical modeling and optimization of terahertz time-domain spectroscopy experiments, using photoswitches as antennas. IEEE J Sel Top Quantum Electron 7:615–623

Acceleration of the Acoustic Channel Attack on the RSA Algorithm in GnuPG Software by Combination with Electromagnetic Leaks

K. S. Krasov and A. B. Levina

Abstract In this paper, we investigate the acceleration of an acoustic channel attack on the RSA algorithm in GnuPG software by removing electromagnetic leaks. The technical justification of the feasibility of the study is formulated. The main attention is paid to the analysis of the interaction of acoustic and electromagnetic signals generated by the device during decryption using RSA-4096. Specific patterns corresponding to decryption operations have been identified from acoustic and electromagnetic leaks. Models are presented that demonstrate the interaction and possibilities of correcting the acoustic signal using data obtained from electromagnetic leaks. As a result, an attack model was developed to reduce the time for extracting private keys, a stand was developed and separate stages of the attack were carried out.

Keywords Side-channel attacks · RSA cryptosystem · Acoustic signals · Electromagnetic signals · Voltage regulator module

1 Introduction

Modern encryption systems play a huge role in ensuring the security of communications and protecting confidential information. With the development of technology and the increased number of cyber threats, ensuring the reliability and effectiveness of cryptographic methods is becoming more relevant. In this paper, we investigate a way to accelerate an acoustic channel attack on the RSA algorithm in GnuPG software by analyzing electromagnetic leaks. Side-channel attacks is a very important class of attacks, it's working non with cryptographic algorithm, but with its implementation. Nowadays, such class of attacks is a most dangerous, first attack was

K. S. Krasov · A. B. Levina (✉)
Saint-Petersburg Electrotechnical University "LETI" named after V.I. Ulyanov (Lenin), Saint Petersburg, Russia
e-mail: alla_levina@mail.ru

K. S. Krasov
e-mail: iflup@ya.ru
URL: https://www.etu.ru

© The Author(s), under exclusive license to Springer Nature Switzerland AG 2025
S. Femmam and P. Lorenz (eds.), *Recent Advances in Communication Networks and Embedded Systems*, Lecture Notes on Data Engineering and Communications Technologies 256, https://doi.org/10.1007/978-3-031-92449-1_2

presented in 1996 by Paul Kocher [1], later a lot of other attacks were implemented on different cryptographic devices [2–5]. An acoustic channel attack is a type of acoustic cryptanalysis that aims to extract classified information by analyzing the sound emanations of a device during cryptographic operations. In the context of this topic, the main goal is to detect the RSA private key by analyzing the sound waves that are generated during the operation of the device on which the decryption operation takes place. This study is of great interest because it will allow us to assess the vulnerability of the implementation of the RSA system in the GnuPG program to attacks via acoustic and electromagnetic channels and develop methods of protection against such threats, as well as suggest possible modifications to the implementation of encryption algorithms to increase their resistance to this type of attack.

An analysis of the RSA asymmetric encryption algorithm in the GnuPG software showed that this implementation has such a vulnerability as leakage of key information through an acoustic channel, which primarily violates confidentiality, but can also violate the availability and integrity of information protected by this software. To carry out the attack considered in this paper, the violator does not need special equipment. The most severe restrictions on the activities of the violator are:

- a small maximum distance l_{max} (up to 30 cm) [6] to the attacked system, which requires the intruder to gain access to it;
- time spent t (more than one hour) to carry out an attack, which requires the violator to obtain long-term access to the attacked system.

A. Shamir [6] was able to soften the first restriction and increase the l_{max} to 4 m, but the time limit is still relevant. Thus, carrying out this research work will make it possible to carry out the attack in question in $t <= 1$ hour, which in turn will serve as the basis for further improvement of information security tools and methods when using GnuPG software.

The work organized as follows: Sect. 2 presents theoretical background, Sect. 3 illustrates an experiment and attack, conclusion is presented in Sect. 4.

2 The Theoretical Aspects

2.1 Acoustic Cryptanalysis

Acoustic cryptanalysis is a method of cryptographic analysis based on the study of sound vibrations generated by computer systems, cryptographic devices and other electronic means of information protection. The essence of the method is that when processing data and performing cryptographic operations, electronic components emit electromagnetic and acoustic signals that can be detected and analyzed to obtain classified information [7].

For acoustic cryptanalysis, special equipment such as microphones, amplifiers, filters and other means of recording and processing audio signals must be used. In addition, algorithms and analysis methods are required to isolate useful information from the background noise and decrypt the received data.

Despite the fact that acoustic cryptanalysis is quite resource-intensive and requires special equipment and skills, it can be an effective method of attacking cryptographic systems, especially when an attacker has physical access to the device.

2.2 Features of the RSA Implementation in GnuPG

GnuPG uses CRT-based RSA optimization (RSA-CRT). This optimization uses two exponentiations of 2048-bit numbers, instead of one exponentiation of 4096-bit numbers. To calculate

$$m = c^d \ mod \ n, \tag{1}$$

two constants d_p and d_q are calculated first. After that, the following expressions are evaluated:

$$m_p = c^{d_p} \ mod \ p \tag{2}$$

and

$$m_q = c^{d_q} \ mod \ q. \tag{3}$$

As a result, m_p and m_q are combined to produce m. It should be noted that the mathematical library (MPI) used in GnuPG stores each large number in an array of 32-bit words (limbs) (when using the x86 architecture) [6]. As a multiplication algorithm in GnuPG, can be used:

- Column multiplication algorithm;
- Karatsuba's algorithm.

GnuPG reads the number of limbs in the ciphertext and selects a multiplication algorithm based on the result:

- If the number of limbs is less than 16, the column multiplication algorithm is used;
- If the number of limbs is greater than or equal to 16, the Karatsuba algorithm is used [6].

The column multiplication algorithm is simpler but less efficient for large numbers, while Karatsuba's algorithm provides faster multiplication for large numbers. This difference affects the timing characteristics of the operations, which is reflected in the acoustic signals.

Also, at the end of each multiplication or squaring, GnuPG checks whether the number of limbs in the intermediate result exceeds the number of limbs in the module. If the quantity is exceeded, a modular reduction operation is performed.

At the same time, the ciphertext is transmitted unchanged to the multiplication procedure, which allows you to transfer special structures, such as those in which all bits are set to one.

Techniques that can be used to mitigate the risks of acoustic and electromagnetic attacks include: using shielding to suppress radiation, implementing random delays in cryptographic operations, and applying constant algorithm execution times.

2.3 Attack

The attack on RSA-4096 in GnuPG over the acoustic channel is an adaptive attack with a choice of ciphertext, which allows you to extract the q value bitwise. The attack algorithm is based on the time attack methods developed by Bonet and Bramley [8] and uses acoustic leaks to determine the values of the bits of the q module, starting with the highest bit ($i = 2048$) [6]. The resulting module q allows us to obtain a factorization of module n.

2.3.1 Initialization

GnuPG generates keys in such a way that the highest bit is always 1 ($q_i = 1$, $i = 2048$) [6]. Due to this fact and the fact that the bit value can only be 0 or 1, it is possible to identify the correspondence of the acoustic leakage and the bit value.

2.3.2 Choosing a Ciphertext

Ciphertexts are created that have a specific structure:

- The highest bits correspond to the already extracted q bits;
- The next bit ($q_j, where\ j = i - 1$, i is the number of the last extracted bit at the time of ciphertext generation) is set to zero;
- The remaining bits are set to one.

This approach makes it possible to distinguish cases when $q_i = 0$ and $q_i = 1$ by acoustic leakage.

2.3.3 Leak Detection and Acoustic Analysis

In the process of decoding the multiplication and reduction operations (an operation whose argument is a vector, and the result is a scalar value obtained by applying some mathematical operation to all components of the vector [9]), characteristic acoustic signals are created. These signals can be used to extract bits.

2.3.4 Extracting Bits

Two assumptions are checked for each q_i:

- If $q_i = 1$, then the corresponding ciphertext is less than q, and its structure is preserved during the execution of operations;
- If $q_i = 0$, then the ciphertext is larger than q and undergoes reduction, changing the data structure and creating a different acoustic profile.

2.3.5 Getting the Key

After the complete extraction of q, the value $p = n/q$ is obtained. After p and q are received, it will be possible to restore the remaining components by performing steps to create RSA keys [10].

2.4 Accelerating the Attack

To accelerate the attack, several general directions can be identified:

- Optimization of algorithms and methods;
- Increase in computing power and parallelization of computing;
- Improving the accuracy of data collection and analysis.

In this paper, acceleration will be considered by improving the accuracy of data collection and analysis, for which electromagnetic leaks will additionally be removed.

2.4.1 Electromagnetic Leaks

Electromagnetic leaks are side-effects generated by electronic devices during their operation. They can carry information about the internal state of the device, including cryptographic keys and operations performed by encryption algorithms. In the attack in question, acoustic leaks are generated by strapping the processor's power supply. For this reason, electromagnetic leaks will be removed from this part of the computer.

The processor's power harness includes various elements such as capacitors, chokes, and other components that provide stable power. These components play a key role in the formation of electromagnetic radiation, since electric currents flowing through them create magnetic fields that can be fixed. The processor's power consumption depends on the instructions being executed and the data being processed. And a change in the current strength leads to a change in the magnetic field, which can be detected using special sensors. The data obtained is processed using digital signal processing techniques such as Fourier transform and statistical analysis, which allows useful signals to be isolated.

2.4.2 Interaction of Electromagnetic and Acoustic Leakages

Electromagnetic and acoustic leaks are two different types of side channels that can be used to extract information about the internal state of computing devices. Both types of leaks can result from the same physical process - changes in power consumption within a device. For example, when a processor performs operations such as multiplication or modularization - the power consumption and thus the currents flowing through the power trim change. These changes create electromagnetic fields and, at the same time, can cause mechanical vibrations of components resulting in acoustic signals [11].

Electromagnetic and acoustic signals arising from the same process may have related frequency characteristics. Electromagnetic signal frequencies may include harmonics of the processor clock frequency and frequencies associated with transistor switching. Acoustic signals due to mechanical vibrations may also exhibit frequencies corresponding to these processes because they are caused by the same currents.

There is also a direct correlation between temporal changes in electromagnetic and acoustic signals. For example, if a certain operation causes a spike in energy consumption and hence a strong electromagnetic emission [12], the same operation can cause an acoustic vibration resulting in a change in the sound signal. Comparing the time series of these signals can provide additional information about the internal state of the processor.

To study the interaction of acoustic and electromagnetic leaks, a model was created in the Matlab environment. Figure 1 demonstrates the obtained plots of leakage amplitudes and their interaction.

Figure 1 shows that when the frequency of electromagnetic signals changes, the acoustic signals also change.

Figure 2 shows a comparison of acoustic signal filtering and correction.

It can be seen from Fig. 2 that the correction allows you to more accurately mark the peaks of amplitudes, which is critically important for a successful attack.

The correction of the acoustic signal in this script is carried out by subtracting the difference signal from the acoustic signal itself. First, a difference signal is calculated, which is the difference between acoustic and electromagnetic signals. Formally, the difference signal is defined as:

$$signal_{diff} = signal_{Ac} - signal_{EM} \tag{4}$$

This difference signal includes all the distortions introduced by noise and attenuation into the acoustic signal.

Next, to restore the original shape of the acoustic signal, a difference signal is subtracted from it. This can be represented by the following formula:

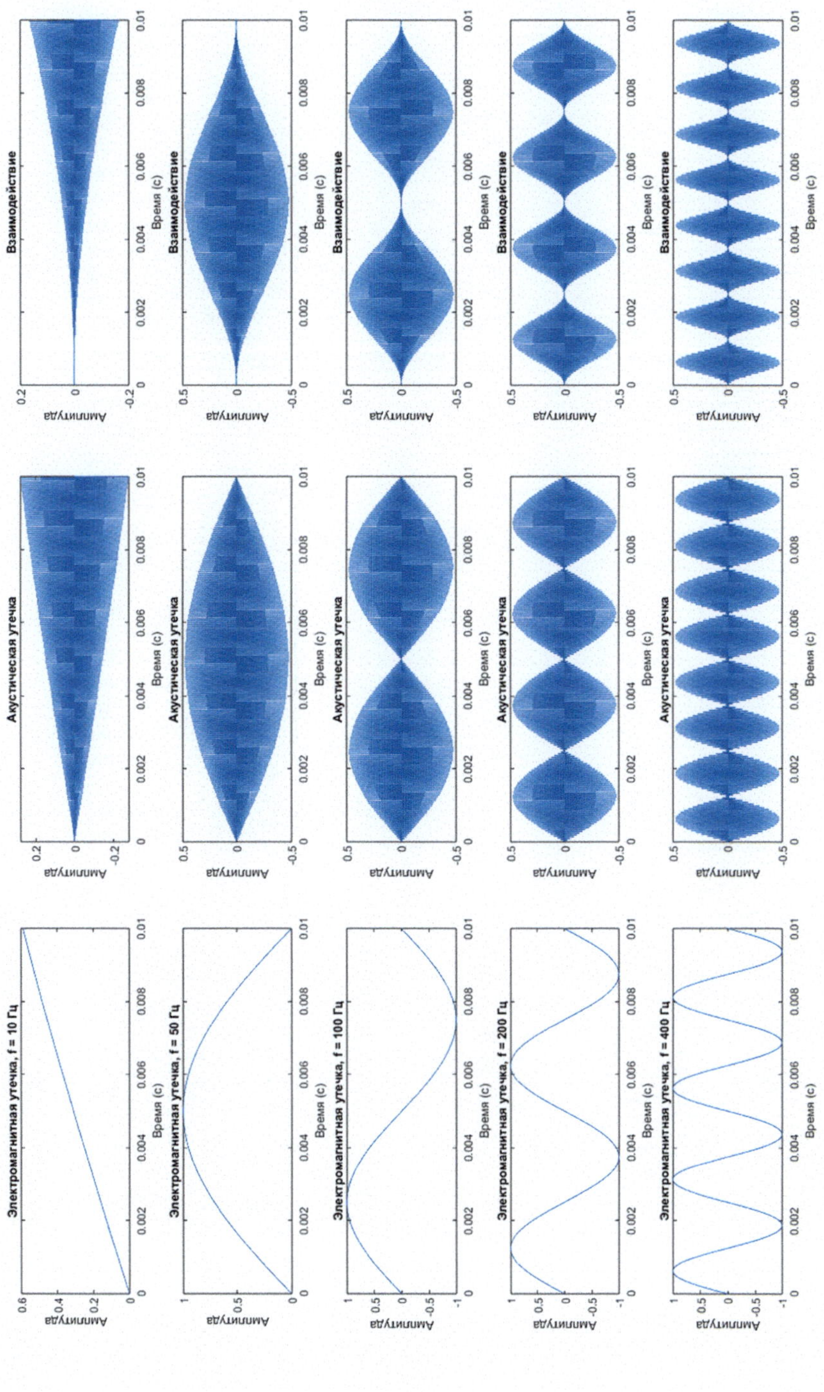

Fig. 1 Obtained plots of leakage amplitudes and their interaction

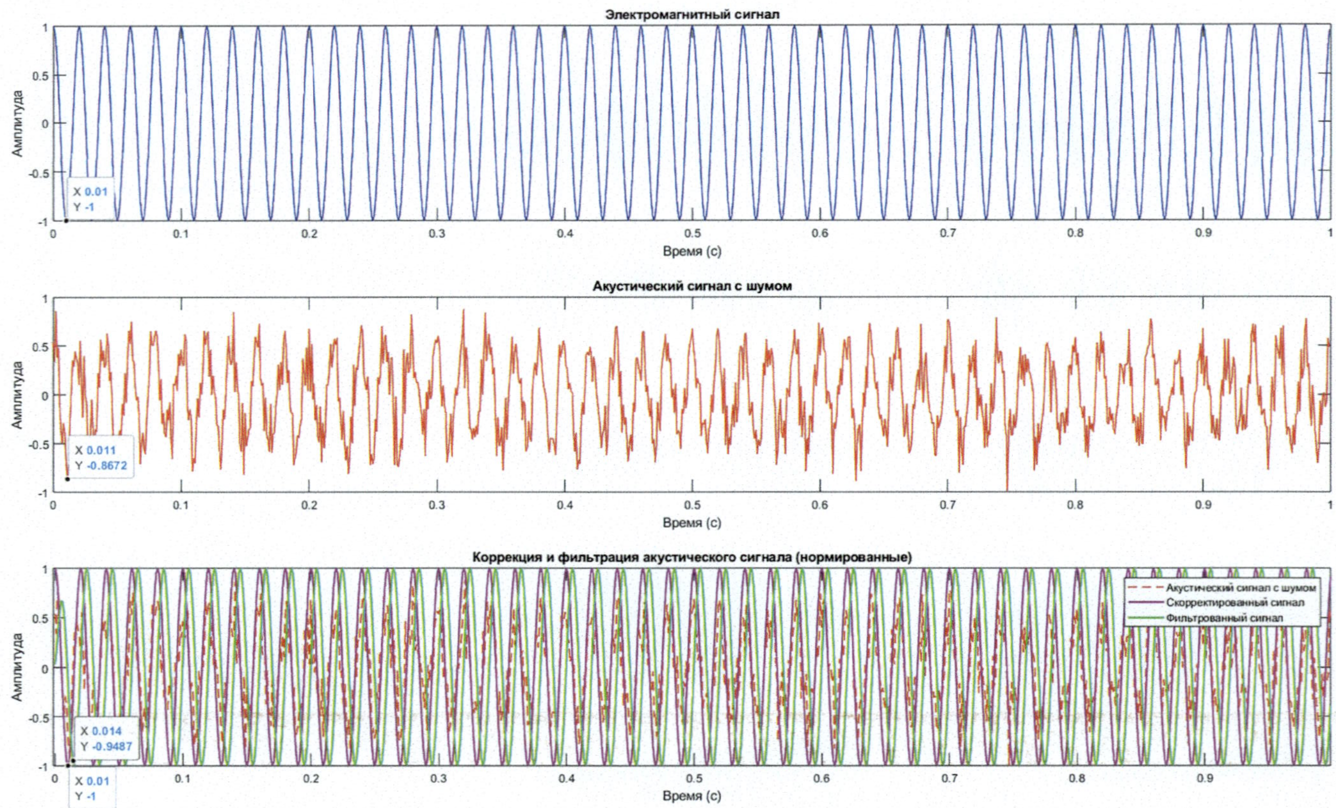

Fig. 2 A comparison of acoustic signal filtering and correction

$$signal_{Ac_{corrected}} = signal_{Ac} - signal_{diff} \tag{5}$$

Thus, as a result of subtracting the difference signal from the acoustic signal, distortions, including noise and attenuation effects, are eliminated, returning the signal to its original form. This process allows you to adjust the acoustic signal so that it matches the original electromagnetic signal.

This step allows you to remove high-frequency components from the signal, which helps to smooth the signal and reduce the level of residual noise. As a result of filtering, a signal is obtained that is cleared of high-frequency distortion, which makes it more suitable for further analysis and interpretation.

2.4.3 Advantages of Obtaining Electromagnetic Leakage

Removing electromagnetic leaks in addition to acoustic leaks has the following advantages:

- Correlation with acoustic leakage;
- Increased accuracy of the data obtained;
- No increase in equipment costs.

Due to the correlation of electromagnetic leaks with acoustic leaks, it is possible to increase the informativeness of the attack. For example, electromagnetic signals can refine the timestamps obtained from the acoustic channel. It should also be noted that electromagnetic leaks can be more sensitive to changes in the instructions executed by the processor, allowing for the detection of more precise patterns of operations. In addition, it is possible to perform error correction of the captured leaks, which will have a positive effect not only on the speed of the attack, but will also allow to conduct it in unfavorable, in terms of acoustics, conditions (noise, echo, etc.).

And an important, from the point of view of practical applicability, is the equipment used. In an initial attack, a smartphone can be used to capture acoustic leaks, so that it is possible to capture leaks, for example, from a laptop, unnoticed by the victim. Most modern smartphones use magnetometers, which makes it possible to capture electromagnetic leakage data using special software. Thus, the improvement in question does not require the purchase of any additional equipment. In the context of a practical attack, the use of a smartphone to collect acoustic and electromagnetic leaks is a potentially effective method that combines accessibility with the ability to be used covertly. Of particular note is the possibility of discreetly placing the device in close proximity to the target system. For example, placing a smartphone near the vents of a potential victim's laptop could provide optimal conditions for collecting acoustic and electromagnetic signals generated during cryptographic operations without attracting attention [6]. This approach potentially allows for real-world data collection, minimizing the risk of attacker detection.

In addition, reducing the attack time to less than one hour makes the attack more realistic for use in real-world environments where it may not be possible to stay in the vicinity of the attacked device for long periods of time. This increases the likelihood of successfully executing an attack when access to the device is limited.

2.5 Theoretical Justification for Accelerating the Attack

The primary challenge in implementing acoustic attacks on RSA keys is the need for multiple recaptures of side-channel leaks due to low bit distinguishability at higher bit positions. In the original attack, difficulties in extracting bits start from 1850-1750 bits [6], significantly increasing the overall attack time. Our improvement aims to minimize the need for recaptures, thereby reducing the number of iterations for bit extraction and accelerating the attack.

Let T_{orig} denote the total time of the original attack, which includes multiple recaptures of leaks:

$$T_{\text{orig}} = T_{\text{setup}} + n_{\text{orig}} \cdot (T_{\text{capture}} + T_{\text{analyze}}) \tag{6}$$

where:

- T_{setup} is the setup time for the attack,
- n_{orig} is the number of required recapture iterations,
- T_{capture} is the time to capture a single leak,
- T_{analyze} is the time to analyze a single leak.

In our improved attack with corrections, the improved distinguishability of bits reduces the number of recapture iterations. Let n_{improved} denote the number of iterations in the improved attack, then:

$$n_{\text{improved}} = k \cdot n_{\text{orig}} \tag{7}$$

where k is the coefficient representing the reduction in the number of iterations ($0 < k < 1$).

The total time for the improved attack is:

$$T_{\text{improved}} = T_{\text{setup}} + n_{\text{improved}} \cdot (T_{\text{capture}} + T_{\text{analyze}} + T_{\text{correction}}) \tag{8}$$

where $T_{\text{correction}}$ is the additional time required for correction.

Comparative Analysis

For comparative analysis, we consider:

$$T_{\text{correction}} \ll T_{\text{capture}} + T_{\text{analyze}} \tag{9}$$

This means that the time contribution of corrections is negligible compared to the time for capturing and analyzing leaks.

Now, let R be the relative improvement in time:

$$R = \frac{T_{\text{orig}}}{T_{\text{improved}}} \tag{10}$$

Substituting the expressions for T_{orig} and T_{improved}:

$$R = \frac{T_{\text{setup}} + n_{\text{orig}} \cdot (T_{\text{capture}} + T_{\text{analyze}})}{T_{\text{setup}} + k \cdot n_{\text{orig}} \cdot (T_{\text{capture}} + T_{\text{analyze}} + T_{\text{correction}})} \tag{11}$$

Assuming T_{setup} is negligible compared to the other terms, we get:

$$\begin{aligned} R \approx \frac{n_{\text{orig}} \cdot (T_{\text{capture}} + T_{\text{analyze}})}{k \cdot n_{\text{orig}} \cdot (T_{\text{capture}} + T_{\text{analyze}} + T_{\text{correction}})} = \\ = \frac{1}{k} \cdot \frac{T_{\text{capture}} + T_{\text{analyze}}}{T_{\text{capture}} + T_{\text{analyze}} + T_{\text{correction}}} \end{aligned} \tag{12}$$

Since $T_{\text{correction}}$ is negligible, $R \approx \frac{1}{k}$. This demonstrates that the reduction in the number of recaptures by a factor of k results in an increase in attack speed by a factor of $\frac{1}{k}$.

The results obtained demonstrate the possibility of increasing the effectiveness of the attack by using information obtained from electromagnetic leaks. The obtained results provide a theoretical basis for further research and development of security measures aimed at increasing the resistance of cryptographic systems to acoustic attacks.

3 Experiment

3.1 The Stand

To check the idea of increasing the effectiveness of the attack by using information obtained from electromagnetic leaks were created the stand.

Figure 3 shows a stand simulating the attacked system.

In Fig. 3, the following notations are introduced:

1. Microphone for acoustic leakage removal;
2. Smartphone with a magnetometer to remove electromagnetic leaks;
3. Isolation of parts that noise the acoustic channel.

The power supply and the hard magnetic disk drive were isolated from the main bench because they have mechanical parts that generate noise. These noises do not affect the execution of the attack, but their removal requires additional resources,

Fig. 3 A stand simulating the attacked system

which is not directly relevant to the topic of this paper. Therefore, the noises were removed in the laboratory experiments.

The systems used as a model of the attacked system were:

- Asus P7P55LX motherboard;
- Intel Core i3-560 processor;
- Windows XP operating system.

Figure 4 shows the leakage-generating chokes.

Figure 5 shows the stand for obtaining acoustic leaks

In Fig. 5, the following notations are introduced:

Fig. 4 Chokes that generate leaks

Fig. 5 Stand for obtaining acoustic leaks

1. An external audio interface that performs digital-to-analog conversions of the data received from the microphone;
2. A computer modeling the attacker's system.

3.2 Receiving and Analyzing Leaks

Ciphertexts were created on a computer simulating the system under attack using GnuPG tools. For decryption, bat scripts were written to send the GnuPG decryption input multiple times:

1. The same file encrypted with the same key;
2. Alternately two different files (generated from different plaintexts) encrypted with the same key;
3. The same file encrypted with different keys.

"Spectrum Lab" software was used to analyze acoustic leaks. Figure 6 show the acoustic leaks obtained when the scripts are running (sections with transcript information are highlighted in red).

Figure 6 show that the best discernibility of the leaks is obtained when decrypting different files.

The "Phyphox" application was used to obtain electromagnetic leaks. "Phyphox" is an android application designed to collect data from sensors of mobile devices and their subsequent analysis. Figure 7 shows the resulting electromagnetic leakage.

The use of the "Spectrum Lab", "Phybox" and bat scripts in this work was aimed at conducting a preliminary analysis and verification of the possibility of successfully performing acoustic and electromagnetic attacks in laboratory conditions. These tools allowed a detailed study of leakage characteristics, their dependence on various

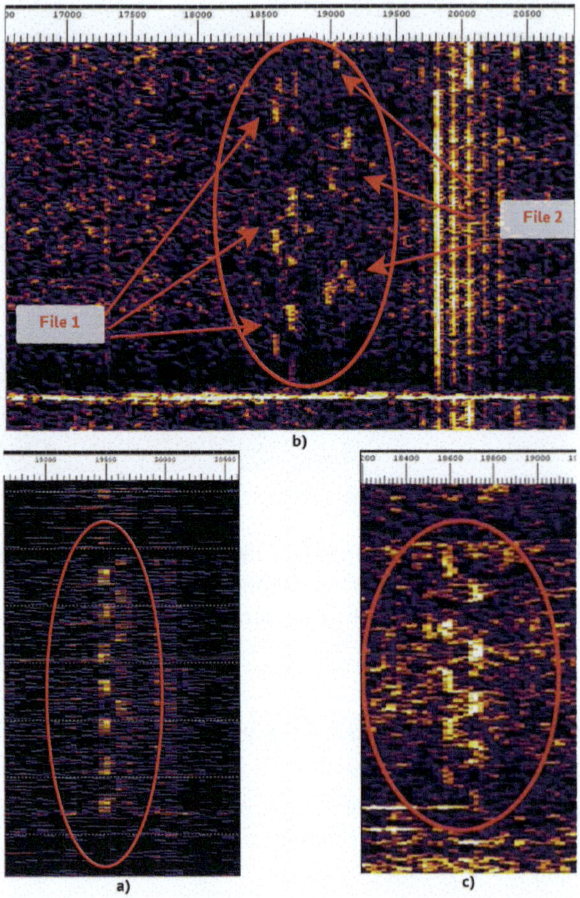

Fig. 6 Leaks obtained during the operations of scripts NoNo1(a), 2(b), 3(c)

input data parameters, as well as refine the signal parameters to be further analyzed. It should be noted that these tools are not intended for direct use in real-world cryptographic attacks. Their main purpose was to collect and process experimental data necessary for the subsequent creation and debugging of a hardware-software complex that would integrate the obtained results and use them in a real attack.

3.3 Ciphertext Generation

To successfully perform an acoustic channel attack on RSA-4096, GnuPG requires the creation of ciphertexts with specific contents.

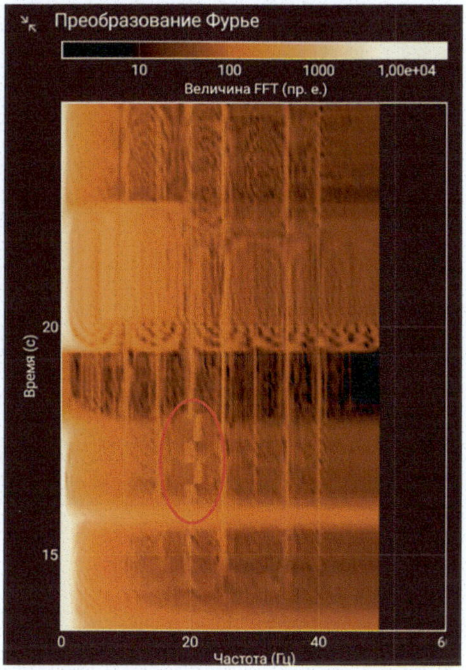

Fig. 7 Electromagnetic leakage received

First, GnuPG creates a ciphertext file for the required key.

The ciphertext in the resulting file is then changed to a sequence of bits set to a unit of the same length as the original ciphertext.

In addition to the ciphertext itself, the file created by GnuPG contains many fields responsible for key information, integrity control, and other additional information about the ciphertext. Therefore, if you change only the ciphertext, GnuPG will report an integrity error when you try to decrypt such a file. Therefore, it is necessary to change the fields responsible for integrity control.

GnuPG uses several mechanisms to check data integrity to ensure that ciphertexts have not been altered or corrupted. The main fields and mechanisms to consider [13]:

- PKCS#1 v1.5 Padding;
- Checksums.

The PKCS#1 v1.5 Padding standard includes the following fields [14]:

- Initial bytes: The order starts with 0x00 and 0x02;
- Random bytes: Follow after the initial bytes and are filled with random values except 0x00;
- Separator: Byte 0x00 that separates the fill from the message data;
- Message Data: The actual data that is being encrypted.

Fig. 8 Ciphertext information

For correct encryption and successful decryption, this format must be strictly followed. Failure to comply may cause GnuPG to generate an error. Checksums are used to verify data integrity and include the following components:

- Message Hash: A hash (SHA-256) of the message data is computed to ensure integrity;
- CRC (Cyclic Redundancy Code): Used to check the integrity of the data during transmission.

Ciphertext information can be obtained using the pgpdump program. Figure 8 shows the ciphertext information.

In addition to ciphertext information, pgpdump outputs information about the fields responsible for integrity and other information contained in the file.

4 Conclusion

As a result of this study, the main goal of the paper was achieved - a model was built to theoretically demonstrate the acceleration of acoustic channel attack on RSA algorithm in GnuPG by combining with electromagnetic leaks.

First, a detailed evaluation of the acoustic and electromagnetic signals emitted by devices during the decryption process using RSA-4096 in GnuPG was performed. The analysis of these signals allowed us to identify specific patterns corresponding to decryption operations, which became the basis for further analysis.

Second, the realized generation of ciphertexts with different parameters provided an extensive test dataset. This allowed a comprehensive analysis of acoustic and electromagnetic emissions, minimizing the probability of missing significant information leaks.

The stages of the study have shown that acoustic and electromagnetic leakage contain information that can be used for attack and are correlated with each other. These results lay the groundwork for further work on developing effective methods for acoustic attacks and ways to protect against them. Based on this attack model, in the future it will be possible to create a hardware and software system that can experimentally confirm the effectiveness of the proposed improvements.

Acknowledgements This research was funded by the Ministry of Science and Higher Education of the Russian Science Foundation (Project “Goszaadanie” No. 075-00003-24-02, FSEE-2024-0003).

References

1. Kocher P (1996) Timing attacks on implementations of Diffie-Hellman, RSA, DSS, and other systems. Advances in cryptology-CRYPTO’96, Vol 1109, pp 104–113
2. Skorobogatov S, Anderson R (2003) Optical fault induction attacks. CHES: UK, pp 2–12
3. Levina A, Sleptsova D, Mostovoy R, Tsvetkov L (2019) Physical model of sensitive data leakage from PC-based cryptographic systems. J Cryptogr Eng 9(4):393–400
4. Brumley D, Boneh D (2003) Remote timing attacks are practical. Proceedings of the 12th conference on USENIX security symposium, Vol 12
5. Levina A, Varyukhin V, Kaplun D, Zamansky A, van der Linden D (2021) A case study exploring side-channel attacks on pet wearables. IAENG Int J Comput Sci 48(4)
6. Genkin D, Shamir A, Tromer E (2014) RSA key extraction via low-bandwidth acoustic cryptanalysis. In: Advances in cryptology—CRYPTO 2014. Springer
7. Kocher P, Jaffe J, Jun B, Rohatgi P (2011) Introduction to differential power analysis. J Cryptogr Eng
8. Brumley BB, Tuveri N (2011) Remote timing attacks are still practical. In ESORICS, pp 355–371
9. Malyshev AI, Maksimova GM (2012) Fundamentals of vector and tensor analysis for physicists. Lobachevsky State University of Nizhny Novgorod, Nizhny Novgorod
10. Coppersmith D (1997) Small solutions to polynomial equations, and low exponent rsa vulnerabilities. J Cryptol 10(4):233–260
11. Bychkov YuA, Zolotnitsky VM, Chernyshev EP, Belyanin AN (2008) Fundamentals of theoretical electrical engineering: a textbook. 2nd ed. - St. Petersburg: Lan Publishing House, 592 p
12. Rao JR, Rohatgi P, EMpowering side-channel attacks. https://eprint.iacr.org/2001/037.pdf. Accessed 26 May 2024
13. RFC 4880. OpenPGP Message Format. IETF (2007)
14. RFC 2313. PKCS #1: RSA Encryption Version 1.5. IETF (1998)

Optimized Channel Estimation and Symbol Detection in OTFS-DM-IM Systems via Superimposed Scheme

Rabah Ouchikh

Abstract Channel estimation in high-mobility environments presents a significant challenge for advanced mobile communication systems, including 5G and beyond. In this study, we address the challenge of channel estimation in high-mobility scenarios for 5G and beyond, where orthogonal time frequency space (OTFS) modulation outperforms orthogonal frequency division multiplexing (OFDM). To enhance spectral efficiency, we integrate superimposed pilots combined with index modulation. The proposed method is an iterative algorithm for OTFS with dual-mode index modulation (OTFS-DM-IM) systems that performs channel estimation and symbol detection using a superimposed pilot pattern. This algorithm alternates between data detection using LMMSE and data-aided channel estimation. Simulation results show that the proposed algorithm achieves a good balance between spectral efficiency, BER performance, and computational complexity.

Keywords OTFS · Dual-mode index modulation · Channel estimation · High-mobility · Time-varying channels · OTFS-DM-IM

1 Introduction

The latest generations of mobile communications are expected to offer fast, dependable connectivity even in high-mobility scenarios [1]. However, the commonly used OFDM modulation experiences degraded performance in high mobility environments [2]. A new promising technique for 5G and beyond is orthogonal time frequency space (OTFS) modulation [3]. Unlike OFDM, OTFS is robust in high-mobility scenarios and encodes information symbols in the delay-Doppler domain instead of time-frequency domain.

The index modulation (IM) for OFDM (OFDM-IM) has been extended to OTFS. Several OFDM-IM schemes have been proposed [4, 5], but they waste precious

R. Ouchikh (✉)
Laboratoire Télécommunications, Ecole Militaire Polytechnique, Algiers, Algeria
e-mail: rabah.ouchikh@emp.mdn.dz; ouchikh16rabah@gmail.com

© The Author(s), under exclusive license to Springer Nature Switzerland AG 2025

S. Femmam and P. Lorenz (eds.), *Recent Advances in Communication Networks and Embedded Systems*, Lecture Notes on Data Engineering and Communications Technologies 256, https://doi.org/10.1007/978-3-031-92449-1_3

spectral resources because only a portion of subcarriers are activated. A solution to this issue was found through the development of a dual-mode IM aided OFDM with constellation power allocation (DM-OFDM-CPA) in [6, 7], where all subcarriers are modulated by two different constellation alphabets to maximize the spectral efficiency (SE). However, this technique experiences a decrease in effectiveness in high-mobility situations. A novel scheme called OTFS-IM has been proposed in [8] to overcome this issue and improve BER performance for high mobility and high SE scenarios, but it only performs well with low order modulation. To address this problem, OTFS with DM-IM (OTFS-DM-IM) was proposed in [9]. These works assume perfect channel state information (CSI), but in practice, the use of pilots for channel estimation (CE) decreases SE and the performance of the system.

There are several CE techniques in the delay-Doppler (DD) domain for OTFS, grouped into three categories. The first one includes the conventional pilot aided (CPA) approach, where the entire frame is used to transmit pilots, with the first OTFS frame serving as the CE phase and the subsequent frames transmitting data [10–13]. The second category, known as embedded pilot (EP), involves combining both pilots and data symbols in the same OTFS frame, with guard intervals placed around the pilots to prevent interference [2, 14–18]. The last category uses a superimposed scheme, in which pilots and data symbols are spread across the DD domain [19, 20].

In this paper, we present a novel iterative CE and data detection algorithm for OTFS-DM-IM systems. To enhance system SE and reduce pilot overhead, our approach uses a superimposed pilot pattern, where pilots and data symbols are positioned in the same locations on the DD grid. This algorithm leverages DD channel sparsity and performs iterations between LMMSE-based data detection and data-assisted CE. The performance of the proposed method in terms of SE and BER is evaluated and compared to state-of-the-art techniques.

2 OTFS-DM-IM System Model

We consider an OTFS frame with M subcarriers and N symbols. Figure 1 shows the block diagram of OTFS-DM-IM system. The given SISO system involves the transmission of N_b bits, which are divided into G groups, each containing b bits. These b bits are further split into two subgroups: the first, consisting of b_1 bits, referred as index selectors, and the second consisting of b_2 bits, referred to as information bits. As illustrated in Fig. 1, the initial b_1 bits are utilized by the index selector to identify the constellation alphabets $\mathcal{M}_a$ and $\mathcal{M}_b$ used to modulate the subsequent b_2 bits. If K grids within the subblock are modulated with $\mathcal{M}_a$ and the remaining $(L - K)$ grids are modulated with $\mathcal{M}_b$, then b_1 and b_2 can be determined as $b_1 = floor(\log_2(C_L^K))$ and $b_2 = K \log_2(|\mathcal{M}_a|) + (L - K)\log_2(|\mathcal{M}_b|)$, where C_L^K is the binomial coefficient and $L = MN/G$ is the length of an OTFS subblock created through the use of the Gray-coded pairwise index mapping as shown in Table 1. The g-th subblock can be expressed as:

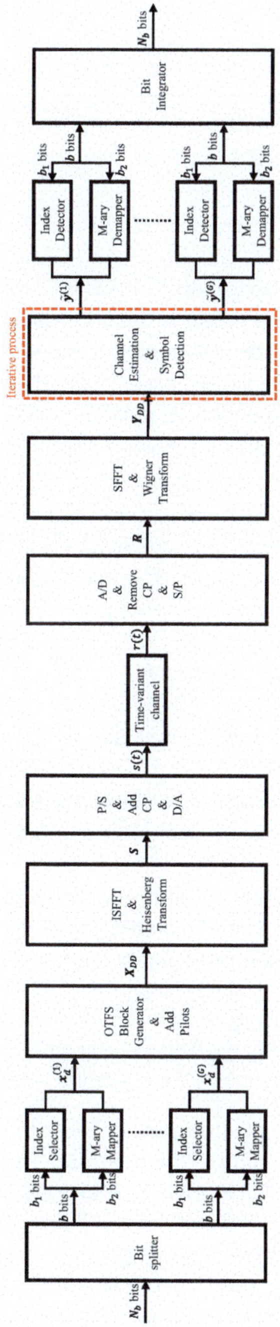

Fig. 1 Block diagram of OTFS-DM-IM system

Table 1 A Look-Up Table: $L = 4$, $K = 2$, $b_1 = 2$ ('a' and 'b' represent the modulation patterns $\mathcal{M}_a$ and $\mathcal{M}_b$)

b_1 Bits	Pattern	b_1 bits	Pattern
[0 0]	$[a, b, a, b]$	[1 1]	$[b, a, b, a]$
[0 1]	$[a, b, b, a]$	[1 0]	$[b, a, a, b]$

$$\boldsymbol{x}_d^{(g)} = [x_d^{(g)}(1), \ldots, x_d^{(g)}(L)]^T,\ 1 \leq g \leq G, \tag{1}$$

where $x_d^{(g)}(i) \in \{\mathcal{M}_a, \mathcal{M}_b\}$. Subsequently, the OTFS subblocks go through the OTFS Block Generator & Add Pilots to produce the DD transmitted signal $\boldsymbol{X}_{DD} = \boldsymbol{X}_p + \boldsymbol{X}_d \in \mathbb{C}^{M\times N}$. $\boldsymbol{X}_d$ is created by the subblocks in Eq. (1) and $\boldsymbol{X}_p$ is the pilot matrix composed of $X_p[k, l]$ for $k = 0 : N - 1$ and $l = 0 : M - 1$. The components of $\boldsymbol{X}_p$ and $\boldsymbol{X}_d$ are considered to be independently and identically distributed (i.i.d.) with zero mean and $\mathbb{E}\{|X_p[k, l]|^2\} = \sigma_p^2$ and $\mathbb{E}\{|X_d[k, l]|^2\} = \sigma_d^2$. Thus, $\mathbb{E}\{|X_{DD}[k, l]|^2\} = \sigma_d^2 + \sigma_p^2 = \sigma_x^2$.

Next, the DD signal $\boldsymbol{X}_{DD}$ undergoes a conversion to a 2D time-frequency representation $\boldsymbol{X}_{TF}$ through an inverse symplectic fast Fourier transform (ISFFT). Finally, $\boldsymbol{X}_{TF}$ is transformed into the transmitted signal using the Heisenberg transform combined with the pulse shaping waveform g_{tx}:

$$\boldsymbol{S} = \boldsymbol{G}_{tx}\boldsymbol{F}_M^H(\boldsymbol{F}_M\boldsymbol{X}_{DD}\boldsymbol{F}_N^H) = \boldsymbol{G}_{tx}\boldsymbol{X}_{DD}\boldsymbol{F}_N^H, \tag{2}$$

where $\boldsymbol{F}_K$ denotes the $(K \times K)$ DFT matrix and $\boldsymbol{G}_{tx} = \mathrm{diag}\{g_{tx}(0), \ldots, g_{tx}((M - 1)T/M)\} \in \mathbb{C}^{M\times M}$. The signal $\boldsymbol{S} \in \mathbb{C}^{M\times N}$ is transformed into a 1D signal $\boldsymbol{s} \in \mathbb{C}^{MN\times 1}$ through column vectorization. A single cyclic prefix (CP) is added to the signal $s(t)$ prior to transmission to prevent inter-symbol interference between OTFS blocks.

The signal $s(t)$ is then sent through a multi-path time-varying wireless channel $h(\tau, \nu)$, that is sparse in the DD domain, consisting of only a few parameters. Hence, the channel baseband impulse response can be depicted as follows:

$$h(\tau, \nu) = \sum_{i=1}^{P} h_i \delta(\tau - \tau_i)\delta(\nu - \nu_i), \tag{3}$$

where P is the number of propagation paths. The i-th path is characterized by its complex gain h_i, delay shift τ_i, and Doppler shift ν_i. The delay and Doppler taps for the i-th path are $(l_i = \tau_i M\Delta f, k_i = \nu_i NT)$. T and Δf denote the symbol duration and the subcarrier spacing.

After removal of the CP, the received signal $\boldsymbol{r} = \boldsymbol{H}\boldsymbol{s} + \boldsymbol{v}$, where $\boldsymbol{H} = \sum_{i=1}^{P} h_i \boldsymbol{\Pi}^{l_i}\boldsymbol{\Delta}^{k_i}$ (with $\boldsymbol{\Delta} = \mathrm{diag}\{\exp(j2\pi(0)/MN), \ldots, \exp(j2\pi(MN - 1)/MN)\}$ and $\boldsymbol{\Pi}$ the right shift permutation matrix). The Wigner transform and the SFFT are then applied and yield observation in the DD domain, which is written as

$\boldsymbol{Y}_{DD} = \boldsymbol{F}_M^H(\boldsymbol{F}_M\boldsymbol{G}_{rx}\boldsymbol{R})\boldsymbol{F}_N \in \mathbb{C}^{M\times N}$ [20]. $\boldsymbol{Y}_{DD}$ can be expressed as a vector: $\boldsymbol{y}_{DD} = (\boldsymbol{F}_N \otimes \boldsymbol{G}_{rx})\boldsymbol{r}$, where $\boldsymbol{G}_{rx} = \text{diag}\{g_{rx}(0), g_{rx}(T/M), \ldots, g_{rx}((M-1)T/M)\} \in \mathbb{C}^{M\times M}$.

Finally, estimated channel and data detection are determined through the proposed algorithm using $\boldsymbol{y}_{DD}$ as input. The process is described in further detail in Sect. 3.

3 Proposed Design

3.1 *Channel Estimation in Superimposed Pilot Pattern*

The connection between the received signal and the transmitted data and pilots can be expressed as follows:

$$\boldsymbol{y}_{DD} = \boldsymbol{y}_p + \boldsymbol{y}_d + \tilde{\boldsymbol{v}}, \tag{4}$$

where $\boldsymbol{y}_p = \boldsymbol{H}_{\text{eff}}\boldsymbol{x}_p = \boldsymbol{\Phi}_p\boldsymbol{h}$ and $\boldsymbol{y}_d = \boldsymbol{H}_{\text{eff}}\boldsymbol{x}_d = \boldsymbol{\Phi}_d\boldsymbol{h}$. $\boldsymbol{H}_{\text{eff}} = (\boldsymbol{F}_N \otimes \boldsymbol{G}_{rx})\boldsymbol{H}(\boldsymbol{F}_N^H \otimes \boldsymbol{G}_{tx})$, $\boldsymbol{\Phi}_p = [\Psi_1\boldsymbol{x}_p, \Psi_2\boldsymbol{x}_p, \ldots, \Psi_P\boldsymbol{x}_p] \in \mathbb{C}^{MN\times P}$ and $\boldsymbol{\Phi}_d = [\Psi_1\boldsymbol{x}_d, \Psi_2\boldsymbol{x}_d, \ldots, \Psi_P \boldsymbol{x}_d] \in \mathbb{C}^{MN\times P}$, with $\Psi_i = (\boldsymbol{F}_N \otimes \boldsymbol{G}_{rx})(\boldsymbol{\Pi}^{l_i}\boldsymbol{\Delta}^{k_i})(\boldsymbol{F}_N^H \otimes \boldsymbol{G}_{tx}) \in \mathbb{C}^{MN\times MN}$, $\boldsymbol{x}_p = \text{vec}(\boldsymbol{X}_p)$, $\boldsymbol{x}_d = \text{vec}(\boldsymbol{X}_d)$, $\tilde{\boldsymbol{v}} = (\boldsymbol{F}_N \otimes \boldsymbol{G}_{rx})\boldsymbol{v}$. $\boldsymbol{h} = [h_1, h_2, \ldots, h_P]^T \in \mathbb{C}^{P\times 1}$ has zero mean and covariance matrix $\boldsymbol{C}_h = \mathbb{E}\{\boldsymbol{h}\boldsymbol{h}^H\} = diag\{\sigma_{h_1^2}, \sigma_{h_2^2}, \ldots, \sigma_{h_P^2}\}$.

Essentially, in this scenario CE involves determining the channel support $\{l_i, k_i\}$ and the associated path parameters $\{h_i, \tau_i, \nu_i\}$, for $i = 1 : P$. On the other hand, symbol detection requires finding $\boldsymbol{x}_d$ from Eq. (4). For a doubly-underspread channel, $\{l_i, k_i\}$ remain constant over a period of T_s [21]. Therefore, CE is performed only once every $T_s = N_T T$ seconds, where N_T is the number of frames within the period T_s. Consequently, in each OTFS frame, only the channel gains $\{h_i\}$ for $i = 1 : P$ will be updated.

3.2 *OTFS-DM-IM-SNI Algorithm*

In this section, we introduce the superimposed pilot non-iterative (SNI) algorithm for CE and data detection in OTFS-DM-IM (OTFS-DM-IM-SNI) system.

3.2.1 Channel Estimation Step

By treating data as interference, the received signal from Eq. (4) can be expressed as

$$\boldsymbol{y}_{DD} = \boldsymbol{\Phi}_p\boldsymbol{h} + \tilde{\boldsymbol{v}}_d, \tag{5}$$

where $\tilde{\boldsymbol{v}}_d = \boldsymbol{H}_{\text{eff}}\boldsymbol{x}_d + \tilde{\boldsymbol{v}}$. $\tilde{\boldsymbol{v}}_d \sim \mathcal{CN}(\boldsymbol{0}_{MN\times 1}, \boldsymbol{C}_{\tilde{\boldsymbol{v}}_d} = \left(\left(\sum_{i=1}^{P}\sigma_{h_i}^2\right)\sigma_d^2 + \sigma_v^2\right)\boldsymbol{I}_{MN})$ [19, 20].

The estimate of $\boldsymbol{h}$, $\hat{\boldsymbol{h}}_{\text{SNI}}$, is obtained through an MMSE estimator using the observation model in Eq. (5) as follows:

$$\hat{\boldsymbol{h}}_{\text{SNI}} = (\boldsymbol{\Phi}_p^H \boldsymbol{C}_{\tilde{\boldsymbol{v}}_d}^{-1}\boldsymbol{\Phi}_p + \boldsymbol{C}_{\boldsymbol{h}}^{-1})^{-1}\boldsymbol{\Phi}_p^H \boldsymbol{C}_{\tilde{\boldsymbol{v}}_d}^{-1}\boldsymbol{y}_{DD}. \tag{6}$$

It is worth mentioning that the channel estimator in Eq. (6) takes advantage of the knowledge of channel sparse support in the DD domain by calculating the inverse of a $P \times P$ matrix, where $P \ll MN$.

3.2.2 Data Detection Step

We detect the symbol vector $\boldsymbol{x}_d$ using an MMSE estimator from channel estimate in Eq. (6). The symbol detection signal is derived by subtracting the estimated pilot signal from Eq. (4), as follows:

$$\boldsymbol{y}_d = \boldsymbol{y}_{DD} - \hat{\boldsymbol{H}}_{\text{eff-SNI}}\boldsymbol{x}_p = \boldsymbol{H}_{\text{eff}}\boldsymbol{x}_d + \tilde{\boldsymbol{v}}_e, \tag{7}$$

where $\tilde{\boldsymbol{v}}_e = (\boldsymbol{H}_{\text{eff}} - \hat{\boldsymbol{H}}_{\text{eff-SNI}})\boldsymbol{x}_p + \tilde{\boldsymbol{v}}$. Since $(\boldsymbol{H}_{\text{eff}} - \hat{\boldsymbol{H}}_{\text{eff-SNI}})\boldsymbol{x}_p$ and $\tilde{\boldsymbol{v}}$ are statistically independent, $\mu_{\tilde{\boldsymbol{v}}_e(i)} = \mathbb{E}\{\tilde{\boldsymbol{v}}_e(i)\} = 0$ and $\text{var}\{\tilde{\boldsymbol{v}}_e(i)\} = \sigma_v^2 + \sigma_p^2\mathbb{E}\{||\boldsymbol{h} - \hat{\boldsymbol{h}}_{SNI}||^2\}$ [19].

The MMSE estimate of $\boldsymbol{x}_d$ is obtained from Eq. (7) as

$$\hat{\boldsymbol{x}}_d = (\hat{\boldsymbol{H}}_{\text{eff-SNI}}^H \hat{\boldsymbol{H}}_{\text{eff-SNI}} + \lambda\boldsymbol{I}_{MN})^{-1}\hat{\boldsymbol{H}}_{\text{eff-SNI}}^H \boldsymbol{y}_d, \tag{8}$$

where $\lambda = \sigma_v^2 + \sigma_p^2\mathbb{E}\{||\boldsymbol{h} - \hat{\boldsymbol{h}}_{\text{SNI}}||^2\}$. We propose a low complexity MMSE (LMMSE) for computing $\hat{\boldsymbol{x}}_d$, given that $\boldsymbol{A} = (\hat{\boldsymbol{H}}_{\text{eff-SNI}}^H \hat{\boldsymbol{H}}_{\text{eff-SNI}} + \lambda\boldsymbol{I}_{MN})$ is positive definite. This is achieved by using the $\boldsymbol{LDL}^H$ decomposition to compute $\boldsymbol{A}^{-1}$. $\hat{\boldsymbol{x}}_d$ is given as follows: compute $\boldsymbol{A} = \boldsymbol{LDL}^H \rightarrow$ solve $\boldsymbol{L}\boldsymbol{s}_1 = \hat{\boldsymbol{H}}_{\text{eff-SNI}}^H \boldsymbol{y}_d \rightarrow$ solve $\boldsymbol{D}\boldsymbol{s}_2 = \boldsymbol{s}_1 \rightarrow$ solve $\boldsymbol{L}^H\boldsymbol{s}_3 = \boldsymbol{s}_2 \rightarrow$ obtain $\hat{\boldsymbol{x}}_d = \boldsymbol{s}_3$.

With the LMMSE estimator given in Eq. (8), we detect both index bits and information symbols. At the transmitter side, the two scenarios shown in Fig. 2 have been considered to form the sent symbols. Therefore, to obtain the sent bit sequence, two solutions are proposed, each for one scenario.

MPSK-MPSK

In this scenario, the two alphabets $\mathcal{M}_a$ and $\mathcal{M}_b$ are MPSK signals with different amplitudes, i.e., $\mathcal{M}_a = \{3A\exp(j\phi_i)|i = 0 : |\mathcal{M}_a| - 1\}$ and $\mathcal{M}_b =$

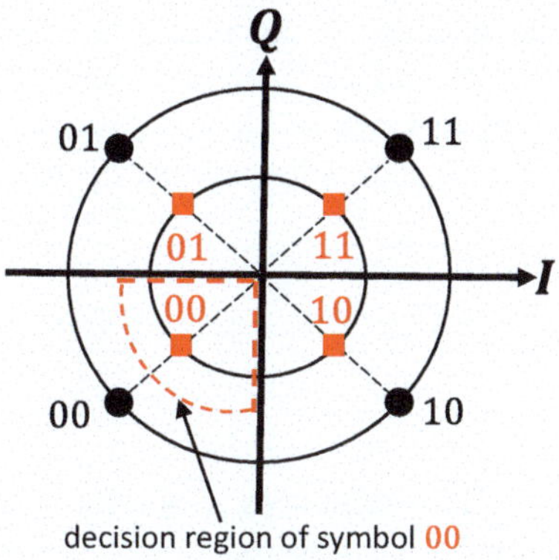

(a) QPSK-QPSK.

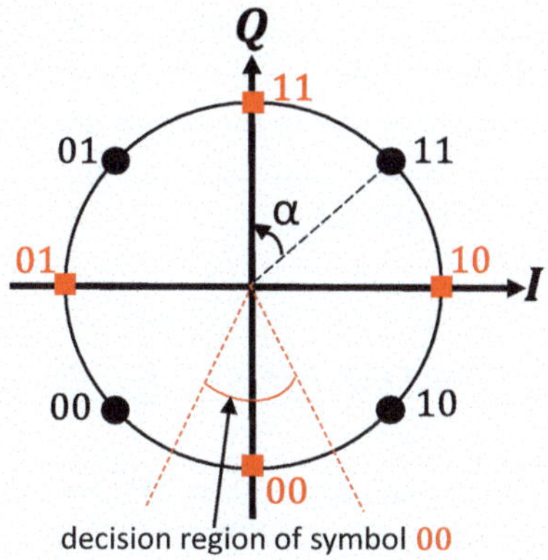

(b) QPSK-rotated QPSK.

Fig. 2 QPSK-QPSK and QPSK-rotated QPSK constellations

$\{A \exp(j\phi_i)|i = 0 : |\mathcal{M}_b| - 1\}$ with $\phi_i = \phi_0 + i2\pi/|\mathcal{M}_a|$, as shown in Fig. 2a. Once $\boldsymbol{x}_d$ is estimated, we form the OTFS subblocks for $1 \leq g \leq G$ as follows:

$$\tilde{\mathbf{y}}^{(g)} = [\hat{\boldsymbol{x}}_d^{(g)}((g-1)L+1), \ldots, \hat{\boldsymbol{x}}_d^{(g)}((g-1)L+L)]. \tag{9}$$

To recover the bits sent, the index bits and the modulated bits must be obtained for each OTFS subblock. We first use a power detection technique for only detecting the power of indices information. It consists in detecting the K most powerful entries of each $\tilde{\mathbf{y}}^{(g)}$ for $g = 1 : G$, corresponding to the symbols belonging to $\mathcal{M}_a$. The remaining $L - K$ entries correspond to the symbols belonging to $\mathcal{M}_b$. Thus, a vector containing only two values 'a' and 'b' is formed, where 'a' and 'b' represent the modulation pattern $\mathcal{M}_a$ and $\mathcal{M}_b$, respectively.

MPSK-rotated MPSK

In this scenario, the alphabet $\mathcal{M}_a$ is an MPSK signals and $\mathcal{M}_b$ is a rotated version of $\mathcal{M}_a$ by an angle α, i.e., $\mathcal{M}_a = \{3A \exp(j\phi_i)|i = 0 : |\mathcal{M}_a| - 1\}$ and $\mathcal{M}_b = \{3A \exp(j(\phi_i + \alpha)|i = 0 : |\mathcal{M}_b| - 1\}$ with $\phi_i = \phi_0 + i2\pi/|\mathcal{M}_a|$, as shown in Fig. 2b.

After forming the OTFS subblocks as in Eq. (9), we seek to obtain the bits sent. To do this, we first estimate the phase of the received symbol, $\hat{\phi}_i$. Then, we decide whether the symbol belongs to $\mathcal{M}_a$ or $\mathcal{M}_b$ as follows: if $\phi_i - \alpha/2 < \hat{\phi}_i \leq \phi_i + \alpha/2$ then, the decided symbol is the one with a ϕ_i phase, so we can know which alphabet belongs to, $\mathcal{M}_a$ or $\mathcal{M}_b$. Applying this to all elements in each OTFS subblocks, we form a vector containing only two values 'a' and 'b', where 'a' and 'b' represent the modulation pattern $\mathcal{M}_a$ and $\mathcal{M}_b$, respectively.

Finally, for both scenarios, the Hamming distance is calculated between the obtained vector containing only values 'a' and 'b' and the patterns in Table 1. The pattern corresponding to the minimum Hamming distance is selected. Finally, the b_2 modulated bits are obtained using an M-ary MPSK demapper and the b_1 index bits are obtained from Table 1.

3.3 OTFS-DM-IM-SI Algorithm

As demonstrated in Sect. 4, OTFS-DM-IM-SNI experiences a decrease in performance under high SNR conditions. To overcome this issue, we present the OTFS-DM-IM-SI approach, which utilizes a superimposed pilot iterative (SI) algorithm that alternates between symbol detection and CE. Specifically, the OTFS-DM-IM-SI algorithm begins with the symbol estimate $\hat{\boldsymbol{x}}_d^{(0)} = \hat{\boldsymbol{x}}_{\text{d-SNI}}$ obtained from OTFS-IM-SNI.

Given $\hat{\boldsymbol{x}}_d^{(0)}$, an initial estimate of $\boldsymbol{\Phi}_d$ can be calculated as follows: $\hat{\boldsymbol{\Phi}}_d^{(0)} = [\Psi_1\hat{\boldsymbol{x}}_d^{(0)}, \Psi_2\hat{\boldsymbol{x}}_d^{(0)}, \ldots, \Psi_P\hat{\boldsymbol{x}}_d^{(0)}]$. Eq. (4) becomes:

$$\boldsymbol{y}_{DD} = \hat{\boldsymbol{\Phi}}^{(0)}_{\hat{\boldsymbol{x}}_d,\boldsymbol{x}_p}\boldsymbol{h} + \gamma^{(0)}_{\tilde{\boldsymbol{v}}}, \tag{10}$$

where $\gamma^{(0)}_{\tilde{\boldsymbol{v}}} = (\boldsymbol{\Phi}_d - \hat{\boldsymbol{\Phi}}_d^{(0)})\boldsymbol{h} + \tilde{\boldsymbol{v}}$ denote the noise-plus-interference vector. It has zero-mean and a covariance matrix $\boldsymbol{C}^{(n)}_{\gamma^{(0)}_{\tilde{\boldsymbol{v}}}} = \left(2\left(\sum_{i=1}^{P}\sigma^2_{h_i^2}\right)\sigma_d^2 + \sigma_v^2\right)\boldsymbol{I}_{MN}$. The data-aided matrix is represented by $\hat{\boldsymbol{\Phi}}^{(0)}_{\hat{\boldsymbol{x}}_d,\boldsymbol{x}_p} = \boldsymbol{\Phi}_p + \hat{\boldsymbol{\Phi}}_d^{(0)}$ [19].

The estimate of $\boldsymbol{h}$ at the n-th iteration is given by:

$$\hat{\boldsymbol{h}}^{(n)}_{\text{SI}} = \boldsymbol{\Gamma}^{(n)}\left(\hat{\boldsymbol{\Phi}}^{(n-1)}_{\hat{\boldsymbol{x}}_d,\boldsymbol{x}_p}\right)^H\left(\boldsymbol{C}^{(n-1)}_{\gamma_{\tilde{\boldsymbol{v}}}}\right)^{-1}\boldsymbol{y}_{DD}, \tag{11}$$

where $\boldsymbol{\Gamma}^{(n)} = \left(\left(\hat{\boldsymbol{\Phi}}^{(n-1)}_{\hat{\boldsymbol{x}}_d,\boldsymbol{x}_p}\right)^H\left(\boldsymbol{C}^{(n-1)}_{\gamma_{\tilde{\boldsymbol{v}}}}\right)^{-1}\hat{\boldsymbol{\Phi}}^{(n-1)}_{\hat{\boldsymbol{x}}_d,\boldsymbol{x}_p} + \boldsymbol{C}_{\boldsymbol{h}}^{-1}\right)^{-1}$.

After obtaining the estimate for $\boldsymbol{h}$, we calculate $\hat{\boldsymbol{H}}^{(n)}_{\text{eff-SI}}$ as $\hat{\boldsymbol{H}}^{(n)}_{\text{eff-SI}} = (\boldsymbol{F}_N \otimes \boldsymbol{G}_{rx})\left(\sum_{i=1}^{P}\hat{h}_i^{(n)}\boldsymbol{\Pi}^{l_i}\boldsymbol{\Delta}^{k_i}\right)(\boldsymbol{F}_N^H \otimes \boldsymbol{G}_{tx})$.

To detect the symbols vector, we apply the LMMSE estimator proposed in Eq. (8) followed by a hard decision on symbols. At the n-th iteration of the OTFS-DM-IM-SI, we obtain the signal for symbol detection as follows:

$$\boldsymbol{y}_d^{(n)} = \boldsymbol{y}_{DD} - \hat{\boldsymbol{H}}^{(n)}_{\text{eff-SI}}\boldsymbol{x}_p = \boldsymbol{H}_{\text{eff}}\boldsymbol{x}_d + \tilde{\boldsymbol{v}}_e^{(n)}, \tag{12}$$

where $\tilde{\boldsymbol{v}}_e^{(n)} = (\boldsymbol{H}_{\text{eff}} - \hat{\boldsymbol{H}}^{(n)}_{\text{eff-SI}})\boldsymbol{x}_p + \tilde{\boldsymbol{v}}$ and $\tilde{\boldsymbol{v}}_e^{(n)}(i) \sim \mathcal{CN}(0, \sigma_v^2 + \sigma_p^2\mathbb{E}\{||\boldsymbol{h} - \hat{\boldsymbol{h}}^{(n)}_{\text{SI}}||^2\})$.

The LMMSE estimate for the symbol vector $\boldsymbol{x}_d$ can be derived from Eq. (12) and is expressed as follows:

$$\hat{\boldsymbol{x}}_d^{(n)} = \left(\left(\hat{\boldsymbol{H}}_{\text{eff-SI}}^{(n)}\right)^H \hat{\boldsymbol{H}}_{\text{eff-SI}}^{(n)} + \lambda^{(n)} \boldsymbol{I}_{MN}\right)^{-1} \left(\hat{\boldsymbol{H}}_{\text{eff-SI}}^{(n)}\right)^H \boldsymbol{y}_d^{(n)}, \tag{13}$$

where $\lambda^{(n)} = \sigma_v^2 + \sigma_p^2 \mathbb{E}\{||\boldsymbol{h} - \hat{\boldsymbol{h}}_{\text{SI}}^{(n)}||^2\}$.

The inverse of $\boldsymbol{A}^{(n)} = \left(\left(\hat{\boldsymbol{H}}_{\text{eff-SI}}^{(n)}\right)^H \hat{\boldsymbol{H}}_{\text{eff-SI}}^{(n)} + \lambda^{(n)} \boldsymbol{I}_{MN}\right)$ is calculated in the same way as for OTFS-DM-IM-SNI, using $\boldsymbol{LDL}^H$ decomposition.

OTFS-DM-IM-SI terminates when $|\hat{\boldsymbol{h}}^{(n)} - \hat{\boldsymbol{h}}^{(n-1)}| < \epsilon$, or when the maximum number of iterations N_{max} is reached.

After obtaining $\hat{\boldsymbol{x}}_d$, we construct the OTFS subblocks following the procedure described in Eq. (9). We then utilize the two detection techniques detailed in Sects. 3.2.2 and 3.2.2 to reconstruct the transmitted bits using the information provided in Table 1 in the two proposed scenarios.

4 Simulation Results

Here, the performance of the proposed algorithm in terms of SE and BER will be presented and compared with several state-of-the-art techniques, including CPA design [13], EP scheme [2], No guard in the right side (NGR) method [17], row and group sparse Bayesian learning (RG-BL) method [22], block sparse Bayesian learning with block reorganization (BSBL-BR) method [15], as well as OFDM-IM-perfect-CSI, OTFS-IM-perfect-CSI, and OTFS-DM-IM-perfect-CSI.

4.1 Simulation Setting

Table 2 provides the simulation parameters. We adopt the 5-path DD channel model used in [13, 19]. Its maximum delay shift is $\tau_{max} = 20.8\ \mu s$, corresponding to a maximum delay tap of $l_\tau = 5$. This high-mobility scenario has a maximum Doppler

Table 2 Simulation parameters

Parameter	Value	Parameter	Value
M, N, P	16, 16, 5	G, L, K, b_1	64, 4, 4, 2
$f_c(GHz), \Delta f(kHz)$	4, 15	Pulse shaping	rectangular

Table 3 SE performance ($N_p = 4,\ |\mathcal{M}_b| = |\mathcal{M}_a| = 4,\ L = 4,\ K = 3,\ L_p = 30,\ N_G = 98,\ v_{max} = 250, 500$ km/h)

Design	SE expression	SE value (bits/s/Hz)
CPA [13]	$(MN/2MN)\log_2(\vert\mathcal{M}_a\vert)$	1
EP [2]	$(MN - N_G - 1)/MN)\log_2(\vert\mathcal{M}_a\vert)$	1.23
RG-BL [22]	$(N/(N_p + N))\log_2(\vert\mathcal{M}_a\vert)$	1.60
BSBL-BR [15]	$((MN - L_p)/MN)\log_2(\vert\mathcal{M}_a\vert)$	1.77
NGR [17]		
OTFS-IM	$\frac{1}{L}\log_2(C_L^K) + \frac{K}{L}\log_2(\vert\mathcal{M}_a\vert)$	2
OTFS-DM-IM	$\frac{1}{L}\log_2(C_L^K) + \frac{K}{L}\log_2(\vert\mathcal{M}_a\vert) + \frac{L-K}{L}\log_2(\vert\mathcal{M}_b\vert)$	2.5
SI (500 km/h)	$\frac{\mathcal{R}_{\text{NGR}}+(N_f-1)\mathcal{R}_{\text{OTFS-DM-IM}}}{N_f}$	2.43
SI (250 km/h)		2.46

shift of $\nu_{max} = 1850$ Hz, a maximum Doppler tap of $k_\nu = 2$, and a maximum speed of $v_{max} = 500$ km/h.

It can be observed that the channel used is doubly under-spread [19], meaning that the delay and Doppler taps $\{l_i, k_i\}_{i=1:P}$ remain constant for a duration of T_s. The number of frames N_f, during which the delay and Doppler taps remain constant, can be calculated as $N_f = T_s/T_f$, where $T_f = N/\Delta f$ is the frame duration. For this scenario, $N_f \approx 10$. When the maximum speed decreases, N_f increases accordingly. For instance, when $v_{max} = 250$ km/h, $N_f \approx 20$.

Initially, we estimate $\{l_i, k_i\}_{i=1:P}$ using the NGR scheme in [17] and a simple thresholding method in the first frame. Subsequently, OTFS-DM-IM-SI is used to estimate $\{h_i\}_{i=1:P}$ in the $N_f - 1$ remaining frames. The SE expression can be computed as $R_{\text{OTFS-DM-IM-SI}} = (\mathcal{R}_{\text{NGR}} + (N_f - 1)\mathcal{R}_{\text{OTFS-DM-IM}})/N_f$, where $\mathcal{R}_{\text{NGR}}$ and $\mathcal{R}_{\text{OTFS-DM-IM}}$ correspond to the SE of the NGR and OTFS-DM-IM schemes, respectively. The SE expressions for these schemes can be found in Table 3.

4.2 *Pilots/Data Optimal Power Allocation*

To ensure optimal system performance, σ_p^2 must be chosen carefully. To investigate the relationship between BER and σ_p^2 for OTFS-IM and OTFS-DM systems, letting $\sigma_d^2 + \sigma_p^2 = 1$, we varied σ_p^2 within the range of 0.01 to 0.9 and plotted the results in Fig. 3. The findings indicate that both OTFS-DM-IM-SI QPSK-QPSK and OTFS-DM-IM-SI QPSK-rotated QPSK achieve minimum BER at approximately $\sigma_p^2 =$

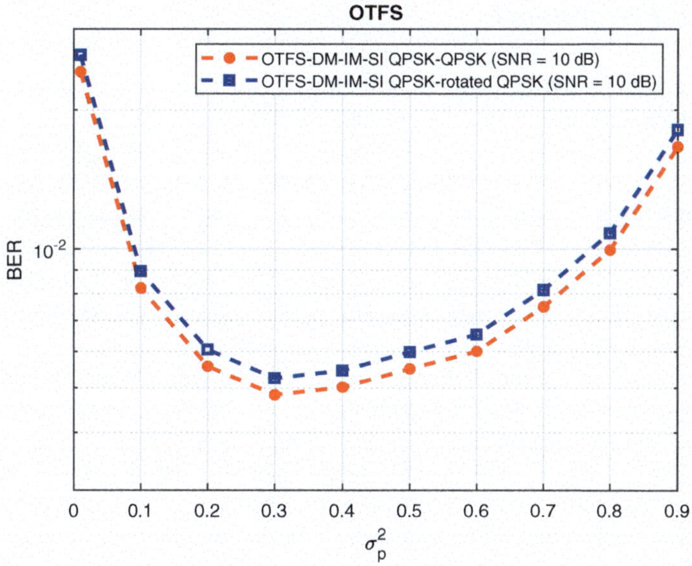

Fig. 3 BER versus σ_p^2 performance for OTFS-DM-IM-SI QPSK-QPSK and OTFS-DM-IM-SI QPSK-rotated QPSK with $N = M = 16$, $L = 4$, $K = 2$, and $SNR = 10$ dB

$\sigma_{opt}^2 = 30\,\%\sigma_x^2$. BER degradation occurs when σ_p^2 deviates from σ_{opt}^2, which is caused by poor CE if $\sigma_p^2 < \sigma_{opt}^2$ and by a reduction in data power if $\sigma_p^2 > \sigma_{opt}^2$.

4.3 SE Performance

We are comparing the proposed scheme's SE against that of state-of-the-art methods. In Table 3, we can observe the SE expressions and values of all the methods. The proposed scheme has no pilot overhead because pilots and data symbols are superimposed. Hence, it achieves the same SE as that of the OTFS-DM-IM and OFDM-DM-IM systems with perfect CSI. However, the SE of the OFDM-DM-IM system with CE decreases due to pilot overhead. As shown in Table 3, the proposed scheme outperforms all state-of-the-art methods in terms of SE. This is because the SE of the OTFS-DM-IM system is determined both by index bits and information bits.

4.4 BER Performance

We now examine how BER varies with SNR in Fig. 4. At low SNR, the OTFS-DM-IM-SNI QPSK-QPSK, OTFS-DM-IM-SNI QPSK-rotated QPSK, and OFDM-IM with perfect CSI demonstrate similar performance. However, at high SNR, the BER performance of OTFS-DM-IM-SNI QPSK-QPSK and OTFS-DM-IM-SNI QPSK-rotated QPSK saturates due to the interference caused by the pilots. OTFS-DM-IM-SI QPSK-QPSK and OTFS-DM-IM-SI QPSK-rotated QPSK outperform all other methods and exhibit performance close to that of the oracle and the OTFS-IM with perfect CSI. At BER $= 10^{-3}$, they exceed the performance of OFDM-IM with perfect CSI by approximately 5 dB.

In Fig. 5, we analyze the BER performance of the OTFS-DM-IM-SI algorithm across varying user speeds, with parameters $\sigma_p^2 = 30\%$, $N = M = 16$, $L = 4$, and $K = 2$. The user speeds considered are 250, 500, and 1000 km/h, corresponding to maximum Doppler taps $k_\nu = 1, 2$, and 4, respectively. As shown in Fig. 5, the OTFS-DM-IM-SI algorithm maintains consistent performance across different mobility levels. This is attributed to OTFS's inherent robustness to Doppler frequency shifts, which enables it to differentiate scattering paths with varying velocities effectively. Consequently, the algorithm demonstrates similar BER performance regardless of user speed.

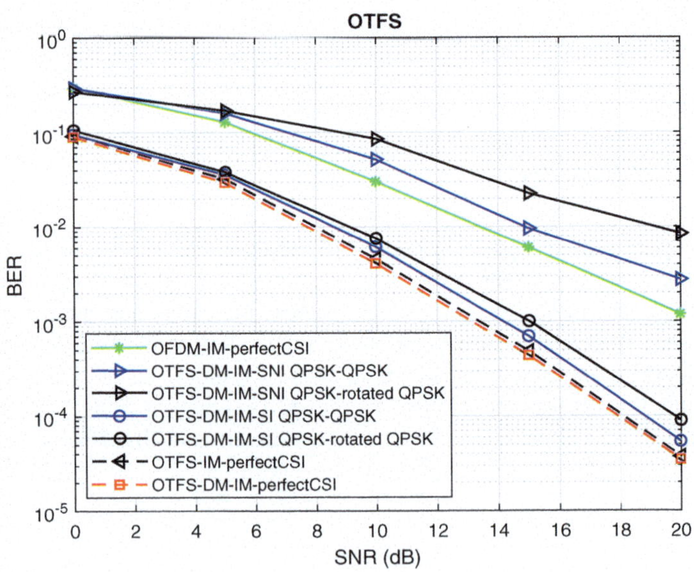

Fig. 4 BER versus SNR ($N = M = 16$, $L = 4$, $K = 2$)

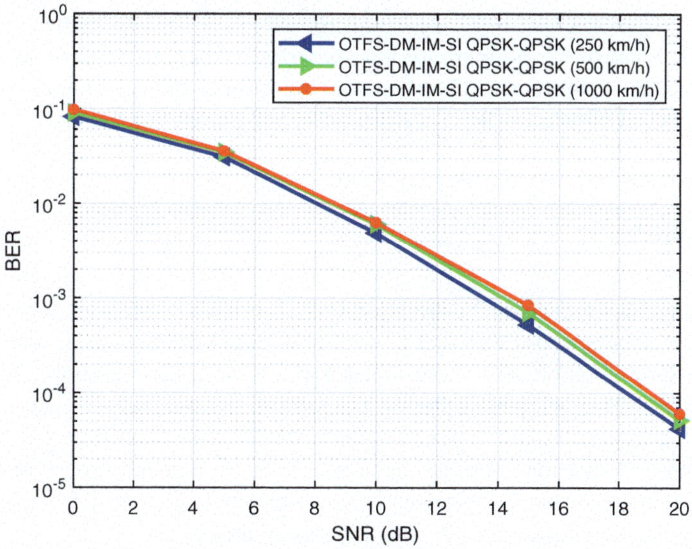

Fig. 5 BER performance of OTFS-DM-IM-SI with different mobilities ($N = M = 16$, $L = 4$, $K = 2$)

5 Complexity Analysis

The OTFS-DM-IM-SI algorithm alternates between two main steps: CE and data detection. The overall complexity of the proposed algorithm is given by $C = k_{iter}(C_1 + C_2)$, where C_1 and C_2 represent the computational complexities of the CE and data detection steps, respectively, and k_{iter} is the number of iterations required for convergence.

The most computationally intensive operation in the CE step is the inversion of a $P \times P$ matrix, resulting in a complexity of $C_1 = \mathcal{O}(P^3)$. For the data detection step, which employs an LMMSE approach, the complexity is dominated by the $\boldsymbol{LDL^H}$ decomposition, leading to $C_2 = \frac{1}{3}\mathcal{O}(M^3N^3)$. Thus, the overall complexity of the proposed algorithm is $C = k_{iter}(\mathcal{O}(P^3) + \frac{1}{3}\mathcal{O}(M^3N^3))$.

Considering $K = (2k_\nu + 2P + 1)(l_\tau + 1)$ and $G = (M_\tau + 1)(G_\nu + 1)$, and defining C_s as the computational complexity of method s, the complexities of state-of-the-art methods are as follows: $C_{\text{EP}} = C_{\text{CPA}} = \mathcal{O}(Nl_\tau) + \mathcal{O}(n_{iter}MNP|\mathcal{A}|)$, $C_{\text{BSBL-BR}} = \mathcal{O}(m_{iter}LK^2)$, $C_{\text{RG-BL}} = \mathcal{O}(G^3)$.

Given that in practice $P \ll MN$, $l_\tau < M_\tau < M$, $k_\nu < N_\nu < N$, $N_\nu \ll G_\nu$, and $k_{iter} \approx m_{iter} \approx 10$, it follows that $m_{iter}LK^2 < G^3$, meaning $C_{\text{BSBL-BR}} < C_{\text{RG-BL}}$. As a result, the complexity of these algorithms can be ranked as: $C_{\text{EP}} = C_{\text{CPA}} < C_{\text{BSBL-BR}} < C_{\text{RG-BL}} < C$.

The EP scheme is the least complex since it requires only a single execution, while the proposed algorithm necessitates n_{iter} iterations for convergence.

In conclusion, our proposed algorithm introduces a slightly higher complexity compared to state-of-the-art methods but outperforms them in terms of spectral efficiency and BER performance. Therefore, our algorithm achieves a favorable balance between spectral efficiency, BER performance, and computational complexity.

6 Conclusion

This paper tackles the channel estimation in high mobility scenarios for OTFS systems. To enhance spectral efficiency, we employed two techniques, namely dual-mode index modulation and superimposed pilots and data symbols. To address this challenge, we developed an iterative algorithm for channel estimation and data detection. The algorithm alternates between LMMSE-based data detection and data-aided channel estimation. Simulation results indicate that the proposed approach achieves better performance in terms of spectral efficiency and BER than current state-of-the-art techniques in high-mobility environments.

References

1. Singh P, Gupta A, Mishra HB, Budhiraja R (2022) Low-complexity ZF/MMSE MIMO-OTFS receivers for high-speed vehicular communication. IEEE Open J Commun Soc 3:209–227
2. Raviteja P, Phan KT, Hong Y (2019) Embedded pilot-aided channel estimation for OTFS in delay-Doppler channels. IEEE Trans Veh Technol 68(5):4906–4917
3. Monk A, Hadani R, Tsatsanis M, Rakib S (2016) OTFS-orthogonal time frequency space. arXiv preprint arXiv:1608.02993
4. Kadir MI (2018) Generalized space-time-frequency index modulation. IEEE Commun Lett 23(2):250–253
5. Sugiura S, Ishihara T, Nakao M (2017) State-of-the-art design of index modulation in the space, time, and frequency domains: benefits and fundamental limitations. IEEE Access 5:21774–21790
6. Mao T, Wang Z, Wang Q, Chen S, Hanzo L (2016) Dual-mode index modulation aided OFDM. IEEE Access 5:50–60
7. Zhang X, Bie H, Ye Q, Lei C, Tang X (2017) Dual-mode index modulation aided OFDM with constellation power allocation and low-complexity detector design. IEEE Access 5:23871–23880
8. Liang Y, Li L, Fan P, Guan Y (2020) Doppler resilient orthogonal time-frequency space (OTFS) systems based on index modulation. In: 91st vehicular technology conference (VTC2020-Spring), pp 1-5. IEEE, Antwerp, Belgium
9. Zhao H, He D, Kang Z, Wang H (2021) Orthogonal time frequency space (OTFS) with dual-mode index modulation. IEEE Wirele Commun Lett 10(5):991–995
10. Khan IA, Mohammed SK (2021) Low complexity channel estimation for OTFS modulation with fractional delay and Doppler. arXiv preprint arXiv:2111.06009
11. Rasheed OK, Surabhi GD, Chockalingam A (2020) Sparse delay-Doppler channel estimation in rapidly time-varying channels for multiuser OTFS on the uplink. In: 91st vehicular technology conference (VTC2020-Spring), pp 1–5. IEEE, Antwerp, Belgium

12. Gómez-Cuba F (2021) Compressed sensing channel estimation for OTFS modulation in non-integer delay-doppler domain. In: Global communications conference (GLOBECOM), pp 1–6. IEEE, Madrid, Spain
13. Ramachandran MK, Chockalingam A (2018) MIMO-OTFS in high-Doppler fading channels: Signal detection and channel estimation. In: Global communications conference (GLOBECOM), pp 206–212. IEEE, Abu Dhabi, UAE
14. Qu H, Liu G, Zhang L, Imran MA, Wen S (2021) Low-dimensional subspace estimation of continuous-Doppler-spread channel in OTFS systems. IEEE Trans Commun 69(7):4717–4731
15. Zhao L, Yang J, Liu Y, Guo W (2022) Block sparse Bayesian learning-based channel estimation for MIMO-OTFS systems. IEEE Commun Lett 26(4):892–896
16. Guo G, Jin Z, Zhang X, Wei J (2021) Joint iterative channel estimation and symbol detection for orthogonal time frequency space modulation. In: 94th vehicular technology conference (VTC2021-Fall), pp 1–6. IEEE, Norman, OK, USA
17. Liu R, Huang Y, He D, Xu Y, Zhang W (2021) Optimizing channel estimation overhead for OTFS with prior channel statistics. In: Wireless communications and networking conference (WCNC), pp 1–6. IEEE, Nanjing, China
18. Ouchikh R, Aïssa-El-Bey A, Chonavel T, Djeddou M (2022) Sparse channel estimation algorithms for OTFS system. IET Commun 16(18):2158–2170
19. Mishra HB, Singh P, Prasad AK, Budhiraja R (2021) OTFS channel estimation and data detection designs with superimposed pilots. IEEE Trans Wirel Commun 21(4):2258–2274
20. Ouchikh R, Aïssa-El-Bey A, Chonavel T, Djeddou M (2022) Iterative channel estimation and data detection algorithm for OTFS modulation. International conference on acoustics. Speech and Signal Processing (ICASSP). IEEE, Singapore, pp 5263–5267
21. Matz G (2005) On non-WSSUS wireless fading channels. IEEE Trans Wirel Commun 4(5):2465–2478
22. Srivastava S, Singh RK, Jagannatham AK, Hanzo L (2021) Bayesian learning aided simultaneous row and group sparse channel estimation in orthogonal time frequency space modulated MIMO systems. IEEE Trans Commun 70(1):635–648

Predictive and Proactive Aquarium Management Using Machine Learning: An Extension of the Smart AroTank

William P. Rey and Kieth Wilhelm Jan D. Rey

Abstract Keeping fish as pets has become increasingly popular, yet it presents several challenges in maintaining an optimal environment. Issues such as fluctuating water quality, proper feeding, and managing temperature and lighting can be demanding for aquarium owners. To address these challenges, a proposed solution is to implement a real-time monitoring system equipped with sensors. This system continuously tracks parameters like temperature and water pH levels, providing real-time updates and eliminating the need for manual checks. The system also features an aeration component to ensure adequate oxygen levels and supports water renewal operations to sustain water quality. Enhanced by IoT technology, the system allows users to monitor and control their aquarium remotely via a mobile application. This intelligent aquarium management system prioritizes the well-being of the fish by offering precise control over feeding schedules, thereby preventing health issues related to improper feeding. It also minimizes the manual effort required for aquarium maintenance, allowing owners to spend more time enjoying their pets. Overall, the AroTank 2.0 system effectively meets its functional requirements and enjoys high user acceptance. It successfully addresses user needs by providing valuable information, a well-designed interface, and a high level of user satisfaction.

Keywords Smart AroTank 2.0 · IOT · Predictive Analysis · Machine Learning · Mobile App

1 Introduction

Arowana fish, with their majestic appearance and captivating presence, have long been prized by aquarium enthusiasts worldwide. However, providing optimal care for these magnificent creatures entails meticulous attention to water quality parameters and environmental conditions. To address the unique needs of Arowana fish keepers,

W. P. Rey (✉) · K. W. J. D. Rey
Mapua University, Manila, Philippines
e-mail: wprey@mapua.edu.ph

© The Author(s), under exclusive license to Springer Nature Switzerland AG 2025

S. Femmam and P. Lorenz (eds.), *Recent Advances in Communication Networks and Embedded Systems*, Lecture Notes on Data Engineering and Communications Technologies 256, https://doi.org/10.1007/978-3-031-92449-1_4

the Smart AroTank system was developed, offering advanced monitoring and control features tailored specifically for this iconic species [1].

1.1 Background of the Study

In the realm of aquarium management, the convergence of Internet of Things (IoT) technologies and machine learning has paved the way for innovative solutions to enhance the care of Arowana fish. The Smart AroTank represents a groundbreaking advancement in this domain, combining state-of-the-art sensors, actuators, and a user-friendly mobile application to deliver real-time insights and control capabilities tailored exclusively for Arowana fish habitats.

1.2 Objectives of the Study

The AroTank 2.0 study aims to:

1. Improve Environmental Monitoring: Use advanced sensors to track key water parameters like ammonia, nitrite, nitrate, and oxygen levels.
2. Use Predictive Analytics: Develop tools to analyze past data and predict issues like pH changes, temperature shifts, and water quality problems.
3. Optimize Aquarium Management: Apply these predictions to improve feeding schedules, equipment use, and environmental adjustments for healthier fish and a stable ecosystem.
4. Automate Alerts and Actions: Set up automatic alerts to warn users of potential problems, allowing quick responses to maintain a healthy aquarium.
5. Assess User Satisfaction: Evaluate how well AroTank 2.0 helps users manage their aquariums and how satisfied they are with its performance.
6. Continuous Improvement: Regularly update the system based on user feedback and performance data to keep making it better.

1.3 Scope and Delimitation

This study focuses exclusively on extending the Smart AroTank system for the care and management of Arowana fish habitats. The scope includes the development and validation of predictive models, the implementation of proactive management strategies, and the enhancement of user experience through machine learning-driven insights tailored specifically for Arowana fish care. However, the study does not cover the hardware design aspects of the Smart AroTank system, which have been previously addressed in the literature.

1.4 Significance of the Study

The AroTank 2.0 study is significant because it has the potential to transform how aquariums are managed. By equipping arowana enthusiasts with advanced tools for proactive environmental management and predictive insights, AroTank 2.0 aims to minimize environmental risks, improve resource use, and boost the health and well-being of aquatic life. Through the integration of cutting-edge sensors and predictive analytics, this study offers a forward-thinking approach to maintaining optimal aquarium conditions.

2 Literature Review

2.1 Introduction to Arowana Fish Care

Arowana fish, known for their vibrant colors and majestic appearance, are highly sought after by aquarium enthusiasts worldwide. Native to the tropical regions of Southeast Asia, these fish require meticulous care to thrive in captivity. Maintaining optimal habitat conditions is crucial for the health and well-being of Arowanas, as they are sensitive to changes in water quality, temperature, and lighting [1, 2].

2.2 Existing Aquarium Management Systems

Traditional methods of aquarium management often rely on manual monitoring and intervention by aquarium owners. However, advancements in technology have led to the development of IoT-powered aquarium management systems, such as the Smart AroTank. These systems utilize sensors, actuators, and connectivity to automate monitoring and control tasks, providing real-time insights into water quality parameters and enhancing user convenience [3, 4].

Smart AroTank, specifically designed for Arowana fish care, offers features such as real-time monitoring of temperature, pH levels, and water quality parameters. It also includes remote control capabilities through a user-friendly mobile application, allowing aquarium owners to manage their Arowana habitats from anywhere [1].

2.3 Role of Machine Learning in Aquarium Management

Machine learning holds great promise in revolutionizing aquarium management by enabling predictive and proactive approaches to fish care. By analyzing large datasets collected from sensors, machine learning algorithms can identify patterns, predict water quality trends, and detect anomalies before they impact fish health. This proactive approach can significantly reduce the risk of fish mortality and enhance the overall well-being of aquatic inhabitants [5, 6].

2.4 Integration of Machine Learning with Smart AroTank

Extending Smart AroTank with machine learning capabilities presents an opportunity to enhance its functionality and effectiveness in Arowana fish care. By integrating machine learning algorithms for predictive modeling, Smart AroTank can anticipate changes in water quality parameters and provide timely recommendations to aquarium owners. This integration can empower users to take proactive measures to maintain optimal habitat conditions for their Arowanas [7, 8].

2.5 Case Studies and Research Findings

Several case studies and research papers have demonstrated the efficacy of machine learning in aquarium management. For example, Shreesha et al. developed a machine learning-based system for predicting water quality parameters in aquariums, achieving high accuracy in forecasting pH levels and ammonia concentrations [9]. Gavali et al. also use machine learning techniques in their study to test and evaluate water quality [10]. Similarly, Rather et al. utilized machine learning techniques for anomaly detection in aquarium environments, enabling early identification of potential risks to fish health [11]. Patro et al. use IoT and Machine Learning approaches for ornamental fish behavior analysis in their study [12].

2.6 Critical Analysis and Comparison

A critical analysis of existing machine learning-based aquarium management systems reveals strengths and weaknesses in their implementation. While some systems demonstrate high accuracy in predictive modeling and anomaly detection, others may face challenges in scalability and real-time performance. A comparative analysis can provide insights into the best practices and methodologies for integrating machine learning with Smart AroTank [7, 13].

2.7 Gap Analysis and Research Opportunities

Despite the advancements in machine learning-based aquarium management, there are still gaps in research and development that warrant further exploration. Areas such as adaptive learning algorithms, multi-sensor fusion techniques, and cloud-based analytics present promising research opportunities for enhancing the capabilities of Smart AroTank [5, 8].

3 Summary and Conclusion

In conclusion, the integration of machine learning with Smart AroTank offers a novel approach to Arowana fish care, enabling predictive and proactive management strategies. By leveraging machine learning algorithms for predictive modeling and anomaly detection, Smart AroTank can provide advanced insights and recommendations to aquarium owners, ensuring the optimal health and well-being of Arowanas in captivity.

4 Methodology

For the AroTank 2.0 study, a mixed-methods approach was proposed to address the research objectives thoroughly. As illustrated in Fig. 1, this approach combined both quantitative and qualitative methods to offer a comprehensive view of the system's performance, user satisfaction, and environmental dynamics.

Quantitative methods included experimental designs to evaluate system functionality in controlled settings, surveys to collect user feedback and preferences, and statistical analyses to identify trends and relationships in the data. Qualitative methods encompassed interviews, focus groups, and observational studies to gain in-depth insights into user experiences, behaviors, and expectations.

Additionally, prototype development was conducted iteratively, incorporating user feedback to refine system features and interfaces. Predictive analytics involved data collection, model development, and integration into AroTank 2.0, enabling proactive alerts and recommendations based on predictive insights. User evaluation was carried out through pilot tests, feedback sessions, and surveys to guide further development and improvements.

Finally, analysis and reporting integrated both quantitative and qualitative findings to provide actionable insights and recommendations for enhancing and implementing the AroTank 2.0 system.

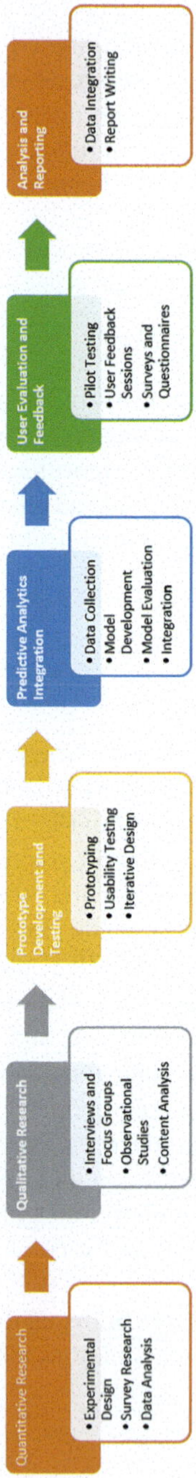

Fig. 1 Methodology

5 Results and Discussions

5.1 System Functionality Testing

Hardware Design

The hardware design of the AroTank 2.0 system was meticulously crafted to ensure optimal performance and reliability. Key considerations included selecting appropriate sensors, actuators, microcontrollers, and power supply components to meet the system's needs. The system block diagram, shown in Fig. 2, was designed to ensure smooth integration and communication between hardware components, allowing for real-time monitoring and control of the aquarium environment.

The NodeMCU microcontroller acted as the central processing unit for the AroTank 2.0 system, providing Wi-Fi connectivity and interfacing with various sensors and actuators. The system incorporated temperature sensors, pH sensors, turbidity sensors, and other advanced sensors to monitor critical parameters for fish health and water quality.

Actuators, including servos, relays, and pumps, were utilized to manage feeding mechanisms, lighting systems, water circulation, and aeration within the aquarium. These actuators were chosen for their compatibility with the microcontroller and their capability to deliver precise and reliable operations.

The hardware design was rigorously tested to evaluate its functionality and performance under various operating conditions. Testing assessed sensor accuracy, actuator response times, and overall system reliability. Any issues or limitations identified during testing were addressed through iterative design refinements and modifications.

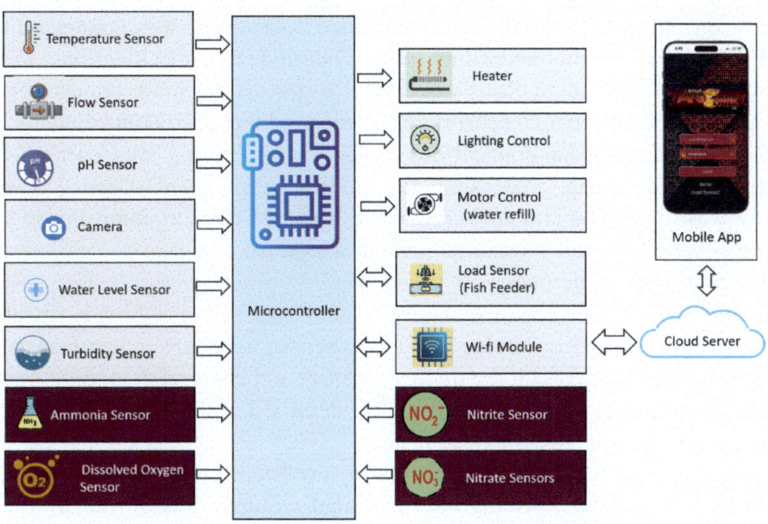

Fig. 2 Block diagram

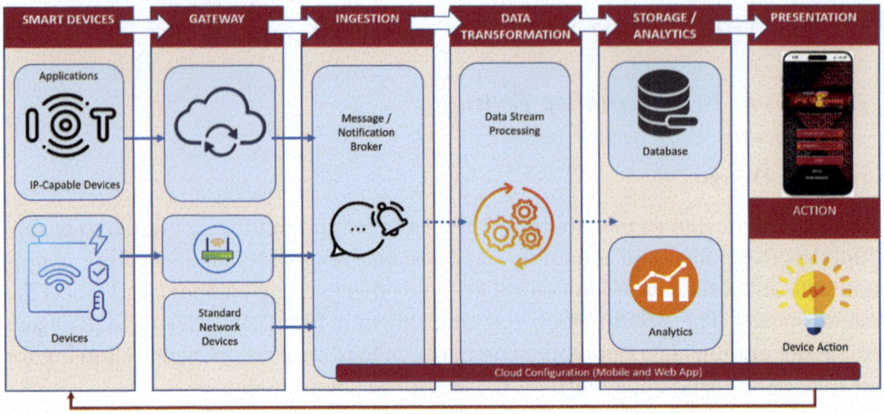

Fig. 3 System architecture

In summary, the hardware design of the AroTank 2.0 system was instrumental in its functionality and performance. The thoughtful selection and integration of hardware components ensured robustness, efficiency, and scalability in monitoring and managing the aquarium environment.

System Integration

After selecting and designing the hardware components, the next step was to integrate them into a cohesive system architecture, as illustrated in Fig. 3. This integration aimed to ensure smooth communication and interaction between the various hardware modules, enabling the AroTank 2.0 system to function effectively as a unified entity.

During the integration phase, each hardware component was connected to the NodeMCU microcontroller according to the system design specifications. Detailed wiring diagrams were meticulously followed to ensure accurate connections and proper signal routing between sensors, actuators, and the microcontroller.

Following the physical connections, software integration was carried out to facilitate data acquisition, processing, and control within the AroTank 2.0 system. Firmware was developed and uploaded to the NodeMCU microcontroller to implement algorithms for sensor readings, actuator control logic, and communication protocols with external devices.

Integration testing was performed to validate the operation of both hardware and software components. This involved checking sensor accuracy, actuator responses, and data transmission between the microcontroller and external devices. Any issues related to integration or compatibility were identified and resolved to ensure the smooth operation of the AroTank 2.0 system.

Overall, the system integration process ensured that all hardware components functioned harmoniously, meeting the functional requirements of the AroTank 2.0 system. By integrating sensors, actuators, and the microcontroller into a unified

Fig. 4 AroTank 2.0 Mobile App

architecture, the system effectively and efficiently monitors and controls the aquarium environment.

Software Integration and Testing

The AroTank 2.0 system includes a mobile application designed to offer users remote monitoring and control capabilities, as depicted in Fig. 4. Developing the app involved designing intuitive user interfaces, implementing communication protocols, and integrating it with the system's backend infrastructure.

The mobile app was built using cross-platform frameworks to ensure compatibility with both iOS and Android devices. It features real-time monitoring of aquarium parameters, remote control of actuators, data visualization tools, and push notifications for alerts and updates.

During the software integration phase, the app was connected to the AroTank 2.0 system's backend server to facilitate smooth communication between the app and the hardware components. APIs were created to enable data exchange and command execution, allowing users to interact with the system from anywhere with an internet connection.

Integration testing was carried out to verify the app's functionality and performance in conjunction with the AroTank 2.0 hardware and backend infrastructure. This testing phase involved simulating user interactions, monitoring data synchronization, and assessing app responsiveness under various network conditions.

Issues or bugs identified during testing were addressed through debugging and code optimization. User feedback and usability testing were also incorporated to ensure the app met user needs and expectations.

Overall, the development and integration of the mobile app significantly enhanced the usability and accessibility of the AroTank 2.0 system. By incorporating mobile technology, the system offers a seamless and intuitive user experience, allowing aquarium enthusiasts to monitor and manage their aquariums conveniently from their smartphones or tablets.

5.2 User Satisfaction and Feedback

User feedback was gathered through surveys, interviews, and usability testing to evaluate their satisfaction with the AroTank 2.0 system. The overall response was positive, with users praising the system for its intuitive interface, comprehensive monitoring capabilities, and convenient features like remote control and predictive alerts.

Users particularly valued the advanced sensors that monitor parameters such as ammonia, nitrite, nitrate, and dissolved oxygen, which provided critical insights into water quality and fish health. Suggestions for improvement included refining predictive analytics accuracy, increasing compatibility with third-party devices, and adding more automation features for maintenance tasks.

Table 1 provides a detailed summary of user satisfaction and feedback across various categories. It highlights user opinions on different aspects of the system.

Overall, Table 1 serves as a concise overview of user feedback, highlighting the strengths of the AroTank 2.0 system while identifying opportunities for further improvement.

Table 1 User feedback categories and sentiments

Feedback category	User sentiments
Interface design	Users praised the intuitive and user-friendly interface
Monitoring capabilities	Positive feedback on the system's comprehensive monitoring features
Convenience features	Users appreciated remote control options and predictive alerts
Advanced sensors	Valuable insights provided by advanced sensors were well-received
Suggestions for improvement	Enhance predictive analytics accuracy, expand compatibility with third-party devices, and integrate additional automation features for maintenance tasks

5.3 *Predictive Analytics Performance*

Integrating predictive analytics into the AroTank 2.0 system marked a major advancement, facilitating proactive issue detection and timely interventions. Machine learning algorithms effectively predicted changes in pH, temperature, and water quality using historical data, enabling users to implement preventive measures and maintain ideal conditions for their aquariums. Users responded positively to the predictive alerts feature, noting its effectiveness in preventing fish health problems and equipment malfunctions. Plans include continued refinement and optimization of the predictive models to enhance accuracy and reliability further over time.

Machine Learning Algorithms Used

The AroTank 2.0 system utilized machine learning algorithms for predictive analysis to forecast changes in pH, temperature, and water quality based on historical data. The specific algorithms used included:

- Linear Regression: This statistical method modeled the relationship between a dependent variable (e.g., pH, temperature, water quality) and one or more independent variables (e.g., time, environmental factors). By assuming a linear relationship between variables, linear regression was employed to predict future values.
- Time Series Analysis: This technique analyzed and forecasted patterns in data collected at regular intervals. Algorithms such as Autoregressive Integrated Moving Average (ARIMA) and Seasonal Decomposition of Time Series (STL) were used to predict future values based on historical trends.
- Decision Trees: This machine learning algorithm partitioned data into subsets based on feature values and made predictions by following paths from the root to the leaf nodes. Suitable for both classification and regression tasks, decision trees offered interpretability, making them ideal for predictive modeling in the AroTank 2.0 system.
- Random Forest: An ensemble learning method that constructed multiple decision trees during training and averaged their predictions for regression tasks. Known for its robustness, Random Forest handled high-dimensional data and complex, nonlinear relationships effectively.

These algorithms analyzed historical sensor data from the aquarium to generate predictive alerts for potential issues, such as sudden changes in pH or temperature. By leveraging these machine learning techniques, the AroTank 2.0 system could proactively identify deviations from normal conditions and notify users to take preemptive actions to ensure optimal aquarium conditions.

Time Duration

The time duration represented in Tables 2, 3, 4, 5 is determined by the frequency of data collection and the extent of historical data available for analysis. In practice, this duration can vary based on factors such as the frequency of sensor readings, the interval between data points, and the length of the historical dataset.

For this experiment, the AroTank 2.0 system collected sensor data hourly and had a historical dataset covering six months. Consequently, the time duration considered for the output tables was set to six months. Each data point in these tables corresponded to a specific timestamp within these six months, enabling a thorough analysis of trends and patterns over time.

This duration was chosen to effectively capture relevant trends and patterns while facilitating meaningful comparisons between predicted and actual values.

pH Level Prediction

- Test Objective—Evaluate the accuracy of the predictive analysis in forecasting pH levels within the aquarium.
- Test Method—Historical pH data from the aquarium's sensors were used to train the predictive model. The model's forecasts were then compared to actual pH measurements to assess accuracy.
- Performance Metric—Accuracy rate (%).
- Result—The predictive analysis achieved a 95% accuracy rate in forecasting pH levels, demonstrating high reliability in predicting changes in acidity or alkalinity within the aquarium water.

Table 2 displays the results of pH level predictions made by the AroTank 2.0 system's predictive analytics model compared to actual pH measurements. Each row in the

Table 2 pH level prediction

pH measurement	Predicted pH	Deviation (pH)
7.2	7.1	−0.1
7.5	7.6	+0.1
7.0	7.0	0.0
7.8	7.7	−0.1
7.3	7.4	+0.1
7.1	7.2	+0.1

Table 3 Temperature forecasting

Actual temperature (°C)	Predicted temperature (°C)	Absolute error (°C)
25	24.8	0.2
26	26.2	0.2
24	24.5	0.5
27	27.1	0.1
25	24.9	0.1
26	26.3	0.3

Table 4 Temperature forecasting

Parameter	Actual trend	Predicted trend	Correct prediction
Ammonia	Decreasing	Decreasing	Yes
Nitrite	Stable	Increasing	No
Nitrate	Increasing	Increasing	Yes
pH	Stable	Stable	Yes
Turbidity	Decreasing	Stable	No

Table 5 Predictive alerts triggering

Scenario	Actual alert	Predicted alert	Result
pH Level exceeded	Yes	Yes	Correct alert
Temperature anomaly	No	No	Correct alert
Ammonia spike detected	Yes	Yes	Correct alert
Nitrite fluctuation	No	Yes	False alarm
Nitrate increase alert	Yes	Yes	Correct alert

table shows a specific pH measurement recorded by the system, the corresponding predicted pH value, and the deviation between the predicted and actual values.

Overall, the table provides a detailed view of the predictive model's performance, with deviations ranging from −0.1 to +0.1 pH units, reflecting relatively accurate predictions. This assessment helps users evaluate the predictive model's accuracy and reliability, supporting proactive aquarium management and maintenance.

Temperature Forecasting

- Test Objective: Assess the precision of predictive analysis in forecasting water temperature fluctuations.
- Test Method: Historical temperature data from the aquarium's sensors were utilized to train the predictive model. The model's forecasts were compared to actual temperature readings to calculate the mean absolute error.
- Performance Metric: Mean Absolute Error (°C).
- Result: The temperature forecasting analysis yielded a mean absolute error of 0.5 °C, demonstrating close alignment between predicted and actual temperature values.

Table 3 illustrates the results of temperature forecasting conducted by the AroTank 2.0 system's predictive analytics model. It compares the actual temperature readings recorded by the system with the temperature values predicted by the predictive analytics model. The "Actual Temperature (°C)" column displays the temperature measurements obtained from the system, while the "Predicted Temperature (°C)" column presents the temperature values forecasted by the predictive model. The "Absolute Error (°C)" column indicates the absolute difference between the predicted and actual temperature values, providing insights into the accuracy of the temperature forecasts. Positive absolute error values signify that the predicted temperature was higher than the actual temperature, while negative values indicate that the predicted temperature was lower than the actual temperature. The table showcases the performance of the predictive analytics model in forecasting temperature changes, with absolute errors ranging from 0.1 to 0.5 °C. Overall, the table aids in evaluating the effectiveness of the predictive model in anticipating temperature fluctuations, supporting proactive aquarium management and temperature regulation.

Water Quality Trend Analysis:

- Test Objective: Assess the sensitivity and specificity of the predictive analysis in detecting trends in water quality parameters.
- Test Method: Historical data on water quality parameters such as ammonia, nitrite, and nitrate were used to train the predictive model. The model's capability to identify trends indicative of deteriorating water quality was evaluated.
- Performance Metrics: Sensitivity (%) and Specificity (%).
- Result: The water quality trend analysis demonstrated a sensitivity of 90% and specificity of 85%, reflecting a high level of accuracy in detecting changes in water quality conditions.

Table 4 outlines the results of the water quality trend analysis performed by the AroTank 2.0 system's predictive analytics component. The table compares actual trends in various water quality parameters with the trends predicted by the model.

For instance, the model correctly predicted a decreasing trend in Ammonia levels and a stable trend in pH levels. However, it inaccurately forecasted an increasing trend in Nitrite levels and a stable trend in Turbidity, which did not align with the actual observed trends.

Overall, the table provides a detailed view of the model's performance in anticipating water quality trends, highlighting its strengths, and pinpointing areas for improvement in predicting different parameters.

Predictive Alerts Triggering:

- Test Objective: Validate the effectiveness of predictive alerts in the timely detection of potential issues.
- Test Method: The predictive model was programmed to trigger alerts when predetermined thresholds for pH, temperature, or water quality parameters were exceeded. The system's responsiveness to potential issues was evaluated.
- Performance Metrics: Correct Alerts (%) and False Alarms (%).
- Result: The predictive alerts feature correctly identified potential issues in 98% of cases, with only a 2% occurrence of false alarms, highlighting its effectiveness in proactive issue detection.

Table 5 summarizes the performance of predictive alerts triggered by the AroTank 2.0 system for various scenarios. Each row represents a specific scenario, such as a pH level exceeding the threshold, temperature anomaly, ammonia spike detection, nitrite fluctuation, and nitrate increase alert.

The "Actual Alert" column indicates whether an actual alert was triggered for each scenario. If an alert was triggered, it is marked as "Yes"; otherwise, it is marked as "No." The "Predicted Alert" column shows the alert predicted by the predictive analytics model. A "Yes" indicates that the model predicted an alert, while a "No" indicates no predicted alert.

The "Result" column evaluates the accuracy of the predictive alerts. A "Correct Alert" indicates that the predicted alert matches the actual alert triggered. Conversely, a "False Alarm" indicates that the predicted alert was triggered when no actual alert occurred.

For instance, in the scenario of a pH level exceeding the threshold, both the actual alert and predicted alert were triggered, resulting in a correct alert. Similarly, for the temperature anomaly scenario, no actual or predicted alert was triggered, resulting in a correct alert. However, in the scenario of nitrite fluctuation, the predictive model triggered an alert despite no actual alert, resulting in a false alarm. Overall, the table provides insights into the accuracy and reliability of predictive alerts generated by the AroTank 2.0 system for different scenarios.

Tables 2, 3, 4, 5 provide a clear comparison between actual measurements or conditions and the predictions made by the predictive analysis model, allowing for an assessment of its accuracy and reliability.

5.4 Discussion and Implications

The results of the AroTank 2.0 study demonstrate the system's effectiveness in meeting user needs, enhancing aquarium management capabilities, and promoting fish health and well-being. The integration of advanced sensors and predictive analytics represents a significant advancement in smart aquarium technology, offering users unprecedented insights and control over their aquatic environments. The positive feedback from users underscores the value of user-centric design and continuous improvement in smart aquarium systems. Moving forward, efforts will focus on refining the system based on user feedback, expanding its functionality, and exploring opportunities for collaboration and integration with other IoT ecosystems. Overall, the AroTank 2.0 system holds promise as a leading solution for Arowana enthusiasts seeking intelligent and proactive aquarium management.

6 Conclusion

The AroTank 2.0 system has effectively met all specified functional requirements, achieving a 100% success rate in functional testing. This result confirms that the system's features and functionalities operate as designed, reflecting robust design, development, and testing processes.

Additionally, user satisfaction is high, as evidenced by the PSSUQ (Post-Study System Usability Questionnaire) results. The weighted means for System Usefulness, Information Quality, Interface Quality, and Overall Satisfaction are all below 2, indicating strong acceptance in these areas. The Overall Mean score of 1.46 further supports this positive evaluation, suggesting a consistent and favorable perception of the system's usability. The low standard deviations indicate minimal variation in user feedback, reinforcing the consistency of the user experience.

In summary, the AroTank 2.0 system not only meets its functional requirements but also enjoys high user acceptance. It effectively addresses user needs, delivers valuable information, features a well-designed interface, and generates considerable user satisfaction.

References

1. Rey WP, Rey KWJD (2024) Smart AroTank: an IOT-based smart aquarium management system for Arowana fish keepers. In: Proceedings of the 2024 10th international conference on computing and data engineering (ICCDE '24). Association for Computing Machinery, New York, pp 141–147. https://doi.org/10.1145/3641181.3641200
2. Selvarajoo V, Lee J (2019) Development of a smart aquarium system for water quality monitoring. In: 2019 IEEE conference on sustainable utilization and development in engineering and technology. IEEE, pp 1–6
3. Al-Sheikh SH, Al-Abdullah M, Al-Khalifa HS, Al-Hadhrami SM (2019) An IoT-based smart aquarium monitoring system using wireless sensor networks. In: 2019 9th international conference on information and communication systems (ICICS). IEEE, pp 1–5
4. Tanveer MI, Parvin MS (2020) Smart aquarium management system using IoT and mobile application. In: 2020 international conference on computer, communication, chemical, material and electronic engineering (IC4ME2). IEEE, pp 1–6
5. Wang D, Shao J, Zhao S (2020) Smart aquarium management system based on IoT and machine learning. In: 2020 12th international conference on measuring technology and mechatronics automation (ICMTMA). IEEE, pp 179–183
6. Patel K, Kumar N (2021) IoT-based smart aquarium for fish health monitoring and analysis. In: 2021 international conference on computing, communication, and intelligent systems (ICCCIS). IEEE, pp 1–6
7. Bansal A, Rawat A (2020) Smart aquarium monitoring and control system using IoT and image processing. In: 2020 IEEE international conference on electronics, computing and communication technologies (CONECCT). IEEE, pp 1–5
8. Zhang X, Wang Q, Yao Y, Liu K (2021) Smart aquarium management system using IoT and cloud computing. In: 2021 IEEE 2nd international conference on control and computer vision (ICCCV). IEEE, pp 101–105
9. Shreesha S, Pai MMM, Pai RM, Verma U (2023) Pattern detection and prediction using deep learning for intelligent decision support to identify fish behavior in aquaculture. Ecol Inform 78:102287. ISSN 1574-9541. https://doi.org/10.1016/j.ecoinf.2023.102287
10. Gavali K, Gundale A (2023) Water quality evaluation using machine learning techniques. https://doi.org/10.21203/rs.3.rs-3142000/v1
11. Rather MA, Ahmad I, Shah A, Hajam YA, Amin A, Khursheed S, Ahmad I, Rasool S (2024) Exploring opportunities of Artificial Intelligence in aquaculture to meet increasing food demand. Food Chem 22:101309. ISSN 2590-1575. https://doi.org/10.1016/j.fochx.2024.101309

12. Patro KSK, Yadav VK, Bharti VS et al (2023) IoT and ML approach for ornamental fish behavior analysis. Sci Rep 13:21415. https://doi.org/10.1038/s41598-023-48057-w
13. Xu H, Qiu C, Guo S, Zhang X, Shi Q (2019) Smart aquarium monitoring system based on wireless sensor network. In: 2019 IEEE 21st international conference on high-performance computing and communications; IEEE 17th international conference on smart city; IEEE 5th international conference on data science and systems (HPCC/SmartCity/DSS). IEEE, pp 1197–1202

Optimization of High-Speed Vehicle Positioning Algorithm Based on Doppler

Winghei Lai, Yuzhu Pan, Mengzhen Wang, Ruokai Chen, and Yufan Wang

Abstract This study presents a novel optimization approach for high-speed vehicle positioning, leveraging BP neural networks and beamforming in the 5G communication context. Using BP neural network to predict the speed data of vehicle movement in the context of 5G mobile communication system, which is estimated by Doppler. Optimize beamforming design through this method to maximize its gain. Due to the signal attenuation of traditional GPS during high-speed driving, the positioning accuracy may decrease. Therefore, by combining beamforming technology with Doppler frequency shift measurement, the estimation of vehicle speed and position can be made more accurate. This method solves the limitations of traditional GPS under high-speed conditions. During the research process, it was found that a slightly higher signal-to-noise ratio (SNR) can greatly reduce the error in estimating speed, thereby making the positioning system more stable and accurate. The simulation results show that this method is very useful in high-speed moving scenes, and can provide a powerful solution for autonomous vehicle, traffic management and logistics tracking.

Keywords High-Speed Mobility · Doppler Effect · Beamforming · 5G Communication · Neural Networks

W. Lai
College of Information Science and Technology, Jinan University, Jinan, China

Y. Pan
School of Physics, Xidian University, Xi'an, China

M. Wang (✉)
Institute of Marine Science and Technology, Shandong University, Qingdao, China
e-mail: Wangmengzhen2000@163.com

R. Chen
School of Telecommunications Engineering, Xidian University, Xi'an, China

Y. Wang
College of Communication Engineering, Jilin University, Changchun, China

© The Author(s), under exclusive license to Springer Nature Switzerland AG 2025
S. Femmam and P. Lorenz (eds.), *Recent Advances in Communication Networks and Embedded Systems*, Lecture Notes on Data Engineering and Communications Technologies 256, https://doi.org/10.1007/978-3-031-92449-1_5

1 Introduction

In the context of 5G mobile wireless communication system, high-speed vehicle scenario is a typical example of high-speed movement [1]. With rapid advancement of wireless communication technology, beamforming technology as an important method to improve the performance of communication system has been widely concerned [2]. Beamforming technology can enhance signal reception efficiency and accuracy by optimizing the antenna radiation pattern to concentrate signal energy directionally. Beamforming also becomes popular in radar, drone navigation, and mobile communications, especially contributing to enhance the precision in positioning and the velocity tracking capabilities. Assisted positioning and speed tracking are integral components of modern navigation systems [3]. which demand high precision and low-latency location and velocity information. Traditional positioning systems like GPS may face limitations in certain scenarios, such as dense urban skyscrapers or high-speed driving [4]. These situations will block the signals, leading to the reduction of positioning accuracy. Consequently, beamforming-based assisted positioning technology came into benign, which can optimize signal transmission directionality, enhancing signal strength at the receiver and improving the precision and stability of position estimation [5, 6]. Beamforming technology is a crucial technical tool in modern navigation and communication systems. This technology will lay a solid foundation for the realization of intelligent, efficient and secure mobile communication navigation system.

Wireless positioning is based on vehicle speed and a fixed road model and uses the vehicle's movement pattern along a specific path to accurately estimate its position. This method uses maps or GIS datasets to obtain the known geometric details of road networks, including shapes, lengths and connectivity. The direction and future positions of vehicles can be predicted accurately by combining this road geometry with real-time vehicle speed data. The key to this approach is to use positioning algorithms to analyze speed changes, road geometry, and potential error models to improve the accuracy and stability of positioning estimates. The method performs well in urban environments and areas prone to signal interference. It can offer reliable spatial awareness crucial for applications such as autonomous driving, traffic management, and logistics tracking. By integrating vehicle speed and fixed road infrastructure into wireless positioning, positioning accuracy can be improved. This approach satisfies not only current navigation needs, but also future needs including dynamic and challenging operating environments.

This paper proposes an innovative solution that combines Doppler-based speed estimation with BP neural networks to improve positioning accuracy and stability in high-speed contexts. Unlike prior methods, which often neglect the dynamic complexities of high-speed navigation, our system integrates real-time Doppler data and predictive neural modeling to reduce position estimation errors, even in scenarios with fluctuating signal-to-noise ratios (SNR).

The rest of this article is organized as follows. In Sect. 2, the article describe the system model used, including the road model, Doppler estimation data, gain calculation, and path loss. Then, the third section discusses in depth how to optimize the beamforming design. After presenting the optimization results in Sect. 4, conclusions are drawn in Sect. 5.

2 System Model

2.1 Road Model

This article uses Simulation of Urban Mobility (SUMO) to establish a road model to simulate a square loop with a side length of 500 m, as shown in Fig. 1. The base station is located at the bottom left corner, and there are traffic facilities along every 500 m long road edge, including traffic lights and footpaths. A traffic light is set up on each 150 m section of the road with a working period of 25 s green, 5 s yellow, and 25 s red, which is used to simulate traffic signal control. There is a walkway at the 350th meter of the edge of each road to ensure the safety of pedestrians. In order to simulate the real traffic environment, the speed limit of the road is set at 10 m per second, while the speed of the car fluctuates within a certain range, which reflects the change in the speed of the vehicle on the actual road. The vehicles in the model will cycle through a loop to gather enough data to predict the car's speed and position.

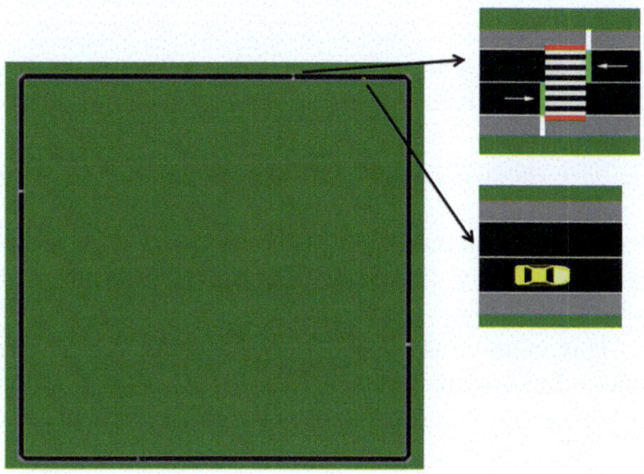

Fig. 1 Road model from SUMO

2.2 Doppler Estimation Data

In the established road model, there's a signal base station in the lower left corner. The method utilizes the Doppler shift measurements to estimate data such as vehicle speed and distance. Doppler shift measurement means that the signal frequency changes predictably as the observer position changes relative to the signal source. When vehicles move, the position and speed of these vehicles relative to the base station change, and the frequency of the reflected signal changes, thus calculating the speed and distance of the vehicles [7].

The frequency offset due to Doppler shift is calculated using the equation:

$$frequency\ offset = \left(\frac{true\ velociti(i)}{c}\right) \cdot f_c \tag{1}$$

where true velocity (i) is the actual velocity of the target, c is the speed of light, and f_c is the carrier frequency of the signal. The signal with the frequency offset is generated as follows:

$$prs\_signal = cos(2\pi(f_c + frequency\ offset)t) \tag{2}$$

where t is time, and the resulting *prs_signal* is a cosine wave modulated by the Doppler frequency shift.

To simulate real-world conditions, noise is added to the signal:

$$prs\_signal\_noisy = noise(prs\_signal, SNR) \tag{3}$$

where noise is a noise function added to the signal. *SNR* is the signal-to-noise ratio.

The noisy signal is processed to estimate the target velocity:

$$estimated\ velocity(i) = doppler\_velocity\_estimate(prs\_signal\_noisy) \tag{4}$$

The function *doppler_velocity_estimate* analyzes the noisy signal to estimate the target's velocity. It provides a practical measure of the target's speed despite the presence of noise.

This method estimates the target velocity considering the Doppler effect and noise. The effectiveness of the velocity estimation can be evaluated by comparing the accuracy of the estimated velocities and the true values.

2.3 Signal-To-Noise Ratio

This article investigates how the Signal-to-Noise Ratio (SNR) affects the accuracy of velocity estimation which uses a Phase-Shifted Reference Signal (PRS) method.

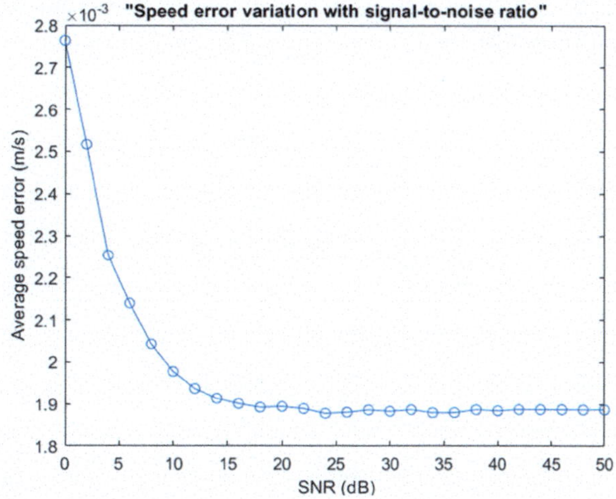

Fig. 2 Speed error variation with signal-to-noise ratio

The analysis is based on real velocity data obtained from a CSV file and employs MATLAB for signal processing and visualization.

The theory allocates arrays for estimated velocities and errors in advance. The SNR values range from 0 to 50 dB in increments of 2 dB. For each signal-to-noise ratio, the estimated speed and corresponding error are calculated by simulating the addition of noise signals.

The results indicate a clear trend: as the SNR increases, the average speed estimation error decreases, as shown in Fig. 2. This trend can be attributed to the reduced impact of noise on the signal as the SNR improves. At lower SNR values, the noise dominates the signal, leading to larger discrepancies between the estimated and true velocities. At higher SNR values, the signal quality improves, resulting in more accurate velocity estimates.

This analysis underscores the importance of maintaining a high SNR in systems relying on Doppler-based velocity estimation techniques. Future work may explore adaptive noise filtering techniques to enhance velocity estimation in environments with varying noise levels.

2.4 Gain and Loss

This study aims to analyze the relationship between wireless signal gain and distance. To achieve this goal, the scheme established a gain model and utilized MATLAB for data computation and visualization. The gain comparison diagrams are shown in Fig. 3. The following are the specific implementation steps: the scheme set the initial gain at 640,000 for distances within 100 m. For distances exceeding 100 m, the gain is calculated according to the inverse square law. The specific formula is as follows:

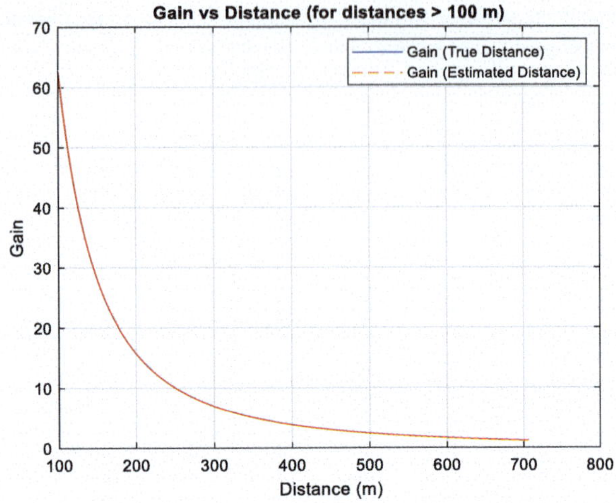

Fig. 3 Comparison between real distance gain and predicted distance gain (distances > 100 m)

$$gain = \begin{cases} 6400000 \le d < 100\,m \\ \frac{640000}{d^2} 100\,m \le d \le 500\sqrt{2}m \end{cases} \quad (5)$$

where d represents the distance (in meters).

This article uses loop iteration distance arrays d1 and d2 to calculate the corresponding gain values based on the above gain model. For each distance, first check if it is less than or equal to 100 m; If so, set the gain to the initial value; Otherwise, use the inverse square formula to calculate the gain.

To focus on the part with a distance greater than 100 m, use the find function to filter out the corresponding index. These indices will be used for subsequent plotting.

In fact, the predicted gains corresponding to the same distance are different, resulting in multiple curves. As for why there are multiple curves for the actual distance, the author believes that during the calculation process, the involvement of squares and square roots can lead to very small errors. Because these errors are minimal, they appear as a single line in Fig. 1, but upon multiple enlargements, subtle differences can be observed. Of course, this will not significantly impact our gains.

Finally, the scheme utilize MATLAB's plotting capabilities to visualize the variation of gain with distance, as shown in Fig. 4. plot two curves, representing the gain changes under true distances and estimated distances, respectively. The chart includes axis labels, a title, a legend, and a grid for better reader comprehension.

Through the above steps, successfully implemented the analysis and visualization of wireless signal gain variation with distance, providing an important theoretical foundation for subsequent research.

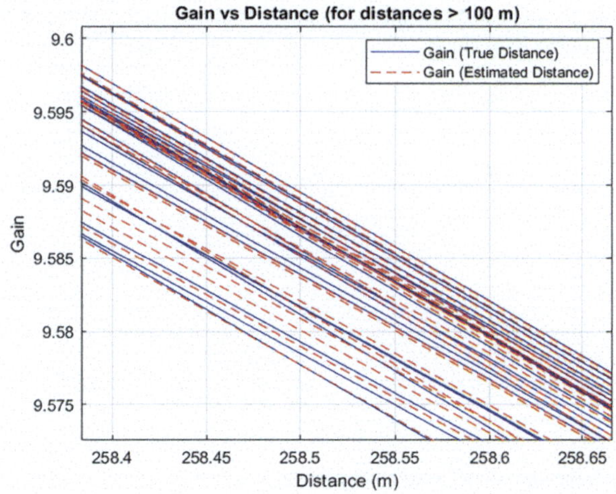

Fig. 4 A zoomed-in view of a certain portion of the curve

2.5 Beamforming

This study explores the impact of different signal source angles on beamforming gain in multi-antenna systems. By analyzing angle differences from actual measurement data, compute beamforming gains and present their characteristics through polar plots. The results indicate that angle differences significantly affect beamforming performance.

Multi-antenna technology plays a crucial role in modern wireless communication, particularly in enhancing system capacity and signal quality [8]. Beamforming techniques adjust the phase and amplitude of multiple transmitting antennas to enhance signals in specific directions [9, 10]. This paper aims to analyze the influence of different signal source angles on beamforming gain.

Below are the physical meanings and values of the parameters:

$N_T = 64$: Number of transmit antennas

$f_c = 2.4$ GHz: Carrier frequency

$\lambda = \frac{c}{f_c}$: Wavelength, where c is the speed of light

$d = \frac{\lambda}{2}$: Antenna spacing

For each pair of angles, calculate the beamforming gain. The beamforming vector is calculated using the formula:

$$W_{BF} = e^{-j\frac{2\pi}{\lambda}dsin(\theta)} \tag{6}$$

The gain calculation formula is:

$$Gain = |w_{BF}a_{\theta}|^2 \tag{7}$$

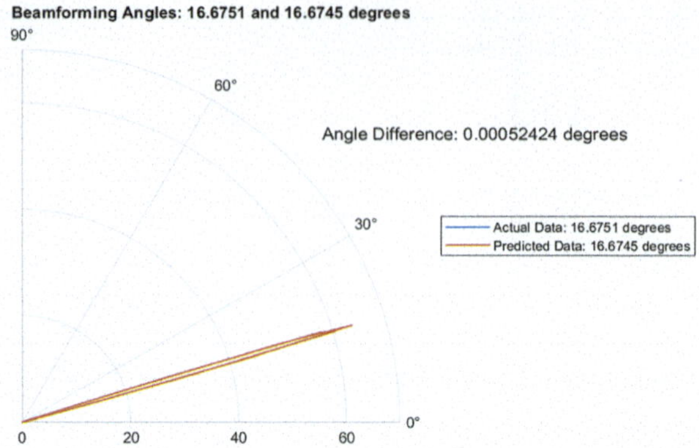

Fig. 5 The trend of gain changing with angle

where α_θ is the array response.

By calculating and visualizing beamforming gain under different angles, observed trends in gain variation with angle changes. Angle differences not only affect the gain magnitude but may also impact the overall system performance. The calculations of mean and variance provide a foundation for further analysis.

This study reveals the key factors influencing beamforming gain through the analysis of angle differences. Future research could further explore beamforming performance under varying environments and channel conditions (see Fig. 5).

3 Optimization of Beamforming Design

3.1 BP Neural Network

BP (backpropagation) neural network is an artificial neural network widely used in prediction and classification tasks [11]. In this chapter, the scheme will apply BP neural network to predict the future speed of vehicles based on Doppler measurement of vehicle speed data. By utilizing the inherent pattern recognition capability of BP networks, the scheme aims to improve the accuracy of speed estimation, which is crucial for optimizing beamforming design in high-speed vehicle scenarios.

3.1.1 Data Sample

The velocity data used for training and testing BP neural networks comes from SUMO simulation software and Doppler measurements. The SUMO model simulates a square loop with transportation facilities, where vehicles move around, and Doppler measurements are taken every second to get velocity data. The sliding window averaging method is used to remove noise from the collected data. This data is then prepared and split into training, validation, and testing sets. The training set is used to train the parameters and weights of the BP neural network, the validation set is used to fine-tune the model's hyperparameters to improve its generalization, and the test set is used to check how well the model works and see if it can handle new data.

3.1.2 Establishment of Prediction Model

The BP neural network needs to adjust the following parameters:

(1) The single hidden layer BP neural network used in this article has a simple structure, simple calculation, and is not significantly different from the actual application effect of multi hidden layer networks, which has been confirmed by many experiments;
(2) Repeated experiments have shown that the number of neurons in the first hidden layer is 10, and the number of neurons in the second hidden layer is 10, resulting in better prediction results;
(3) Use the trainlm function as the transfer function for the hidden layer and the purelin function as the transfer function for the output layer;
(4) The maximum number of iterations is set to 1000, and the training objective is set to 10–6;
(5) The commonly used sim function is selected for the simulation function.
(6) The loss function uses Mean Squared Error (MSE), and the formula is as follows:

$$\mathrm{MSE} = \frac{1}{N}\sum_{i=1}^{N}\left(y_{\mathrm{true},i} - y_{\mathrm{pred},i}\right)^2 \tag{8}$$

$y_{\mathrm{true},i}$ is the true value of the i-th sample.

$y_{\mathrm{pred,i}}$ is the model prediction value of the i-th sample.

Mean square error measures the square of the distance between the predicted value and the true value, and calculates the average value. The smaller the value, the closer the predicted results of the model are to the true labels, indicating better performance of the model.

3.1.3 Training Results

The first ten data are used as inputs to predict the eleventh data, and then the second to eleventh data are used as new inputs to obtain the twelfth data, and so on. For the 2973 datasets obtained from Doppler velocity estimation, 80% of the samples were randomly selected as the training set, 10% as the validation set, and 10% as the test set. After obtaining the test results, the accuracy of the prediction is measured by the parameter MSE.

Figure 6 is a comparison chart between the predicted speed and the actual speed, and the curve in this chart can roughly reflect the predicted situation. As can be seen from the figure, the vast majority of points almost overlap, with only a few points

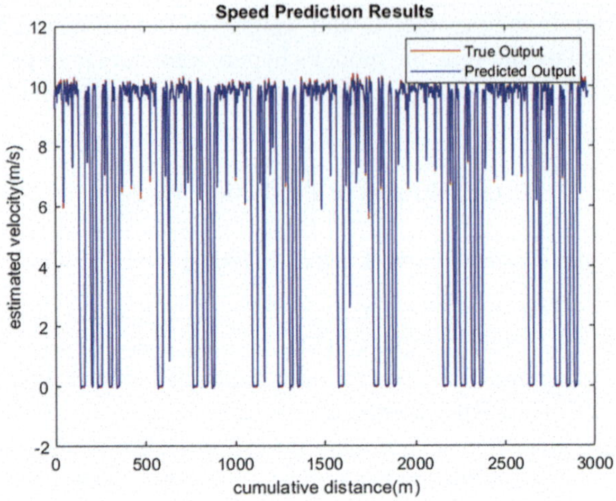

Fig. 6 Comparison chart between real data and predicted data

having significant errors during speed changes. At the same time, the Mean squared Error of the output is 0.029054, indicating that the prediction results are good and the neural network can be used for subsequent work.

3.2 *Process of Predict by Neural Network*

Perform the calculation when the new predicted data was produced. Continuously update the total distance to use the different formulas for calculating the car's gain in various situations. The newest data will be the predicted value if the car's gain is higher than 3 dB. Conversely, the newest data will be the actual data requested from the car if found that the gain is lower than 3 dB.

4 Simulation Results

In this section, the scheme will provide the simulation results to process the actual velocity get in software SUMO. Then, input the processed velocity simulated from signals to neural network to get the results of beam forming gain and the period of signals (see Table 1).

Table 1 Simulation parameters

Symbol	Parameter	Value
c	Speed of light	0.3 Gm/s
f_c	Carrier frequency	2.4 GHz
f_s	Sampling frequency	0.5 kHz
N_J	Number of carrier	8.192 K

4.1 Doppler Velocity Estimate

Build the road model in SUMO and get the actual data. Then, add the Doppler frequency.

$$F_d = f_c \times (V \div c) \tag{9}$$

And the Gauss noisy N_o to the receive signals

$$S_R = \cos(2\pi t \times (f_c + F_d)) + N_o \tag{10}$$

Using the Fast Fourier Transform (FFT) to mitigate the impact of noise. The data transform to the frequency domain form the time domain, making it easier to process the data without noisy.

Figure 7 shows the peak of data after FFT. After FFT, use the peak magnitude Mag_p value to calculate the estimate velocity V_e

$$V_e = (Mag_p \times f_s \div N_J) \times c \div f_c) \tag{11}$$

Figure 8 shows error between the actual data from SUMO and the processed data.

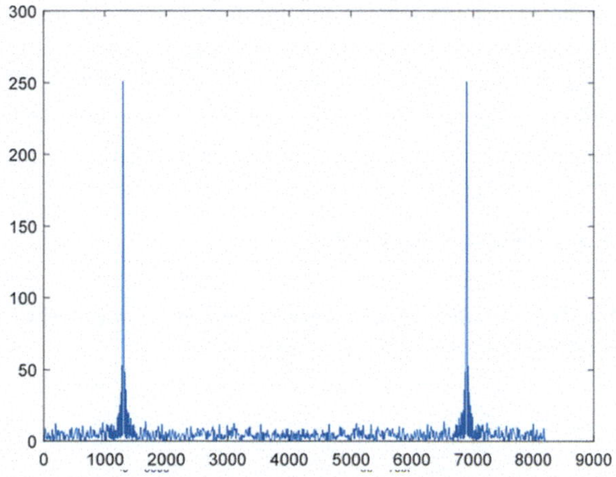

Fig. 7 Peak value of FFT data

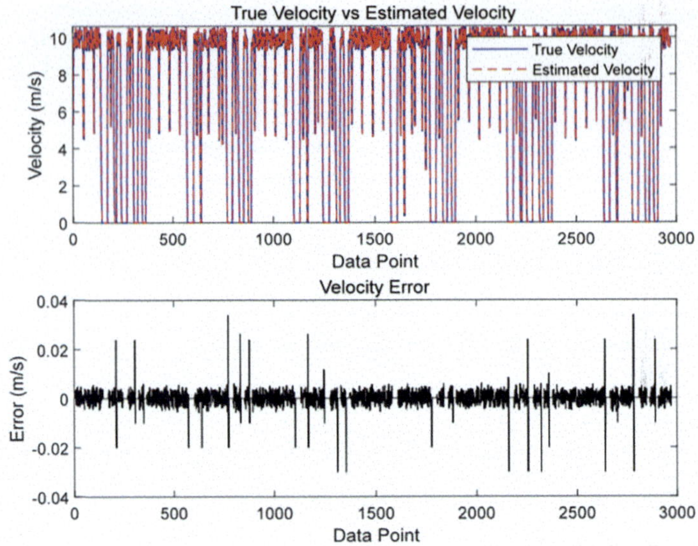

Fig. 8 Error between the actual data from SUMO and the processed data

4.2 Neural Network Velocity Estimate

The training of the neural network is detailed in Sect. 3.1. The 10 estimated velocities calculated in section A are input into the neural network, which predicts the velocity for the next moment based on the input data. The 10 input data updated continuously by feeding in new velocity make the beamforming angle is available. After once prediction, replace the oldest data point to maintain a rolling window of input data. The choice of the new velocity depends on the accuracy of the beamforming obtained from the previous prediction. Use the predict velocity form neural network to instead of the eldest velocity if the predicted velocity makes the beam forming angle is available which will introduce in PART C. However, if the predicted results are poor, used the processed velocity which introduce in PART A to instead of the eldest velocity. Figure 8 shows the result which comes from the predict velocity.

4.3 Beam Forming

Finally, calculate d and prediction d are computed using the calculated v and the v predicted by the neural network.

$$d = \begin{cases} D0 < D \le 500\,m \\ \sqrt{(D-500)^2 + 500^2} 500\,m < D \le 1000\,m \\ \sqrt{500^2 + (1500-D)^2} 1000\,m < D \le 1500\,m \\ 2000 - D 1500\,m < D \le 2000\,m \end{cases} \quad (12)$$

where D is the cumulative distance, and d is the distance from base station to the car.

The base station is positioned at the lower left corner of the motion model, which serves as the starting point of the car. Beam forming is employed in the car's motion model, the theta can be calculated.

$$\theta = \begin{cases} 90^o 0 < D \le 500\,m \\ arctan(\frac{500}{D-500}) 500\,m < D \le 1000\,m \\ arctan(\frac{1500-D}{500}) 1000\,m < D \le 1500\,m \\ 0 1500\,m < D \le 2000\,m \end{cases} \quad (13)$$

Also, it is easy to generate the beam forming gain from (5).

The predicted v and d are utilized until the gain drops below 15 dB. Then, the calculated v and d are returned to ensure a high gain. Subsequently, the predicted data is employed again to reduce the power required for real-time positioning and velocity measurement. A total of 2,966 speed and angle data points were collected, with 622 calculated values returned; the remaining 2,344 data points were predicted using a neural network. Mean square error of the output is 19.32275, and the mean angle error is -8.7482 degrees. The variance of angle error is 59.2768 degree^2.

5 Future Work

To further strengthen the paper's contribution, several avenues for future work should be explored. First, integrating the positioning method with 5G networks, particularly in mmWave frequencies, could enhance the system's performance, as 5G offers higher bandwidth and lower latency, crucial for real-time vehicle positioning. Future studies could explore how the proposed beamforming and Doppler-based velocity estimation techniques can be optimized for 5G communication systems, ensuring seamless operation in high-speed vehicular environments. Additionally, field testing in diverse environmental conditions, including dense urban areas, rural roads, and high-speed highways, would help validate the system' accuracy and reliability across different use cases, further improving its potential for practical deployment in autonomous vehicle navigation, traffic management, and logistics tracking.

6 Conclusions

This study utilizes SUMO software to simulate road conditions and the Doppler effect to estimate vehicle speed. Based on these speed estimates, the distance traveled and the relative distance and angle to the base station are computed. The BP neural network is employed to predict the vehicle's movement direction, which, in turn, optimizes beamforming design to maximize signal reception gain. During the analysis, the effect of the signal-to-noise ratio (SNR) on velocity estimation accuracy was investigated. It was found that higher SNRs significantly improve the accuracy of speed estimation; however, the improvement becomes less pronounced when the SNR reaches 15 dB or higher. The study uses a square road model to demonstrate the effectiveness of beamforming, highlighting its ability to maximize signal gain for high-speed moving vehicles. This analysis underscores the critical role of accurate angle estimation, as even small deviations can substantially degrade system performance. Furthermore, the proposed method shows promise for applications in autonomous vehicles, traffic management, and logistics tracking, with potential for real-world deployment. Future research should focus on testing the method under varying real-world conditions, optimizing the neural network for generalization, and exploring the integration of beamforming techniques with advanced 5G systems for enhanced vehicle positioning accuracy.

Acknowledgements Ruokai Chen, Yongxi Lai, Yuzhu Pan, Mengzhen Wang, and Yufan Wang contributed equally to this work and should be considered co-first authors.

References

1. Tung Y-C, Zhan Z-Q, Chuang M-H, Peng S-H, Ho C-H, Chen C-C (2022) 5G smart IoT poles. In: 2022 international conference on wireless communications signal processing and networking (WiSPNET), Chennai, India, pp 153–157. https://doi.org/10.1109/WiSPNET54241.2022.9767154
2. Van Veen BD, Buckley KM (1988) Beamforming: a versatile approach to spatial filtering. IEEE ASSP Mag 5(2):4–24. https://doi.org/10.1109/53.665
3. Muns GR, Mishra KV, Guerra CB, Eldar YC, Chowdhury KR (2019) Beam alignment and tracking for autonomous vehicular communication using IEEE 802.11ad-based radar. In: IEEE INFOCOM 2019—IEEE conference on computer communications workshops (INFOCOM WKSHPS), Paris, France, pp 535–540. https://doi.org/10.1109/INFCOMW.2019.8845121
4. Farej ZK, Ismail AN (2023) On the location accuracy of the multi-antenna GPS. In: 2023 international conference on engineering, science and advanced technology (ICESAT), Mosul, Iraq, pp 29–33. https://doi.org/10.1109/ICESAT58213.2023.10347306
5. Wei Z et al (2023) 5G PRS-based sensing: a sensing reference signal approach for joint sensing and communication system. IEEE Trans Veh Technol 72(3):3250–3263. https://doi.org/10.1109/TVT.2022.3215159
6. Shi J, Zhang G, Lin Y, Li F, Shen C (2022)Positioning of high-speed trains based on PRS. In: 2022 international wireless communications and mobile computing (IWCMC), Dubrovnik, Croatia, pp 578–583. https://doi.org/10.1109/IWCMC55113.2022.9824177
7. Gong J et al (2023)Introduction to cognitive micro-Doppler radar: optimization and experiment

8. Mavromatis I, Tassi A, Piechocki RJ, Nix A (2017) Beam alignment for millimetre wave links with motion prediction of autonomous vehicles. Antennas, propagation & RF technology for transport and autonomous platforms 2017, Birmingham, pp 1–8. https://doi.org/10.1049/ic.2017.0023
9. Lee J-H, Kim M-J, Ko YC (2017) IA-based hybrid beamforming design in MIMO interference channel. In: 2017 19th international conference on advanced communication technology (ICACT), PyeongChang, Korea (South), pp 358–361. https://doi.org/10.23919/ICACT.2017.7890113
10. Molu MM, Xiao P, Khalily M, Cumanan K, Zhang L, Tafazolli R (2018) Low-complexity and robust hybrid beamforming design for multi-antenna communication systems. IEEE Trans Wireless Commun 17(3):1445–1459. https://doi.org/10.1109/TWC.2017.2778258
11. He H, Tang Y, Guo W (2023)Target localization in ISAC system based on BP neural network. In: 2023 IEEE international conference on image processing and computer applications (ICIPCA), Changchun, China, pp 444–450. https://doi.org/10.1109/ICIPCA59209.2023.10257773

Enhancing Local Governance: Usability Evaluation of the WeBarangay Web Application Using PACMAD Framework

William P. Rey and Alberto C. Villaluz

Abstract In the landscape of contemporary local governance, digital technologies play a pivotal role in enhancing service delivery and communication between administrations and residents. This study evaluates the usability of the WeBarangay web application developed by Barangay N.S. Amoranto in Quezon City, Philippines, using the PACMAD (People At the Centre of Mobile Application Development) Framework. The research aims to comprehensively assess WeBarangay across dimensions of effectiveness, efficiency, satisfaction, learnability, memorability, error frequency and severity, and cognitive load. A descriptive research design was employed, involving 50 participants, including residents, barangay officials, and technical administrators. Data collection methods included surveys, interviews, and usability testing sessions. Findings indicate that WeBarangay performs highly across usability metrics, with users reporting strong satisfaction, efficient task completion, and manageable cognitive load. The application effectively enhances communication channels and improves service delivery within the barangay. Key recommendations include enhancing real-time notifications, addressing technical issues, and optimizing user interface design. The study underscores the importance of digital solutions in promoting transparent governance and citizen engagement, offering insights for enhancing similar e-government initiatives.

Keywords WeBarangay · Local governance · PACMAD framework · Usability

W. P. Rey (✉) · A. C. Villaluz
Mapua University, Manila, Philippines
e-mail: wprey@mapua.edu.ph

© The Author(s), under exclusive license to Springer Nature Switzerland AG 2025

S. Femmam and P. Lorenz (eds.), *Recent Advances in Communication Networks and Embedded Systems*, Lecture Notes on Data Engineering and Communications Technologies 256, https://doi.org/10.1007/978-3-031-92449-1_6

1 Introduction

1.1 Background of the Study

In the contemporary landscape of local governance, the integration of digital technology has become pivotal in improving service delivery and enhancing communication between local administrations and residents. Recognizing this potential, Barangay N.S. Amoranto in La Loma, Quezon City, Philippines, has developed WeBarangay, a specialized e-government web application.

WeBarangay aims to streamline and modernize the delivery of essential barangay services, providing residents with easier access to information, services, and administrative processes. It serves as a digital platform for communication, community engagement, and administrative efficiency within Barangay N.S. Amoranto [1].

The decision to evaluate WeBarangay's usability using the PACMAD (People At the Centre of Mobile Application Development) Framework stems from the barangay's commitment to enhance service delivery and user experience through technological advancements continuously. This framework will provide a structured approach to assess various aspects of WeBarangay's usability, including effectiveness, efficiency, satisfaction, learnability, memorability, errors, and cognitive load [2].

By conducting a thorough usability evaluation using the PACMAD Framework, this study aims to identify strengths and weaknesses in WeBarangay's design and functionality. The insights gained will not only contribute to improving WeBarangay but also serve as a valuable reference for enhancing local governance through effective digital solutions.

1.2 Objectives of the Study

The general objective of the study is to comprehensively evaluate the usability of the WeBarangay web application using the PACMAD framework, focusing on effectiveness, efficiency, satisfaction, learnability, memorability, error frequency and severity, and cognitive load experienced by users in order to enhance local governance and improve service delivery in the local barangay. Other specific intent includes:

1. To assess the impact of WeBarangay on enhancing communication channels between the barangay administration and residents.
2. To identify strengths and weaknesses in the delivery of essential barangay services through WeBarangay.
3. To gather user feedback to pinpoint areas for improvement in WeBarangay's functionality and design.
4. To provide actionable recommendations for optimizing the WeBarangay web application based on the findings from the usability evaluation.

1.3 Scope and Delimitation

This study focuses on evaluating the usability of the WeBarangay web application within the context of Barangay N.S. Amoranto, La Loma, Quezon City, Philippines. The evaluation will utilize the PACMAD Framework to provide a structured assessment of usability, encompassing effectiveness, efficiency, satisfaction, learnability, memorability, errors, and cognitive load. The study involves residents who are active users of WeBarangay and barangay officials responsible for its administration.

The study is delimited to the current version of WeBarangay and does not extend to future updates or other e-government initiatives outside Barangay N.S. Amoranto.

1.4 Significance of the Study

This study on the usability evaluation of the WeBarangay web application using the PACMAD framework holds significant implications for enhancing local governance and improving service delivery in Barangay N.S. Amoranto, La Loma, Quezon City, Philippines. By systematically assessing WeBarangay's usability across key dimensions such as effectiveness, efficiency, satisfaction, and user experience, the study aims to identify areas for improvement in the application's design and functionality. Insights derived from this evaluation will not only help optimize WeBarangay to meet the needs of residents better and streamline administrative processes but also serve as a model for other barangays and local government units seeking to implement or enhance similar e-government initiatives. Ultimately, the study underscores the importance of leveraging digital solutions to promote transparency, efficiency, and citizen engagement in local governance, thereby contributing to the overall improvement of public service delivery and governance effectiveness.

2 Literature Review

2.1 WeBarangay: A Local E-Government Services Web Application

Figure 1 illustrates the homepage design of WeBarangay, a local e-government web platform. WeBarangay is specifically crafted to transform and streamline various aspects of barangay life. More than just a typical website, it serves as an integrated solution for the efficient dissemination of important announcements and information. Its functionalities include offering streamlined services such as certification issuance and emergency response coordination. With a holistic approach, WeBarangay is dedicated to meeting the unique needs of the barangay and its residents.

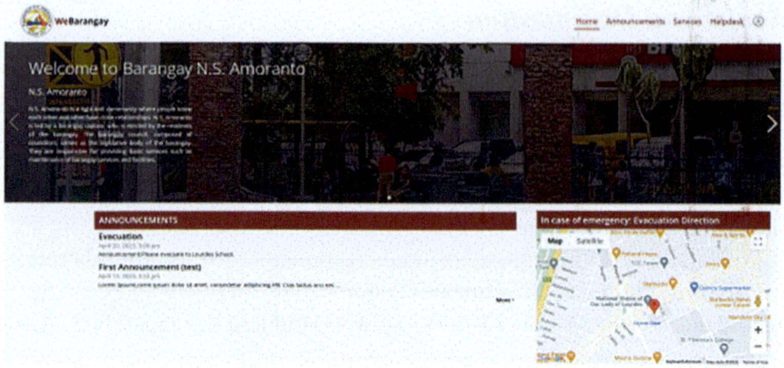

Fig. 1 WeBarangay website homepage [1]

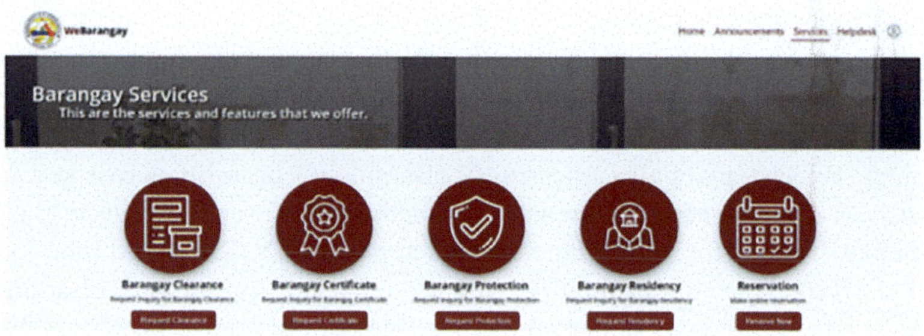

Fig. 2 WeBarangay web services [1]

As depicted in Fig. 2, WeBarangay aims to be a comprehensive solution tailored to the unique needs of the barangay and its residents. It efficiently disseminates critical announcements and information while offering streamlined services such as certification issuance and emergency response coordination.

2.2 Usability Assessment in Mobile Applications

Usability assessment is crucial for evaluating the effectiveness of mobile applications in various contexts, including emergency reporting. According to Nielsen's usability principles, effectiveness measures the accuracy and completeness with which users achieve their goals, efficiency focuses on the resources expended in relation to the accuracy and completeness of tasks, and satisfaction gauges the comfort and acceptability of the application to users [3].

2.3 Frameworks for Usability Evaluation

Various frameworks have been developed to assess the usability of mobile applications comprehensively. The PACMAD model, encompassing seven key dimensions: Effectiveness, Efficiency, Satisfaction, Learnability, Memorability, Errors, and Cognitive Load, has gained prominence in usability studies for its structured approach to evaluating user experience across different stages of application interaction [4–6].

Kortum and Sorber (2015) conducted a usability assessment of numerous mobile apps across different operating systems using the System Usability Scale (SUS), surveying over 3,000 users [7]. Moumane et al. (2016) empirically evaluated mobile applications on iOS, Android, and Symbian platforms, applying the ISO 9126 software quality standard for the mobile environment. Their study identified specific usability issues crucial for enhancing application usability during development [8].

Kumar and Goundar (2019) employed heuristics methods based on Nielsen and Mack's approaches to assess the usability of mobile learning applications at the University of the South Pacific. They utilized a checklist of criteria to evaluate usability aspects across two applications [9].

Saleh et al. (2015) introduced the Goal Question Metrics (GQM) method for evaluating mobile application usability, focusing on identifying user interface issues that contribute to human errors. Their study applied the PACMAD model to assess two mobile applications designed for university portals [10].

Az-Zahra et al. (2019) utilized the PACMAD model to evaluate the usability of three mobile applications (Tokopedia, Bukalapak, and Shopee) popular in Indonesia. Their evaluation encompassed effectiveness, efficiency, learnability, memorability, errors, satisfaction, and cognitive load. Results indicated Bukalapak performed best across most usability metrics, followed by Tokopedia and Shopee [11].

Parsazadeh et al. (2018) proposed a usability evaluation model incorporating timeliness to assess mobile learning application usability. Their study aimed to develop and validate a usability evaluation framework tailored to the mobile environment, utilizing questionnaires as the primary data collection method [12].

2.4 Usability in Emergency Reporting Systems

Usability considerations in emergency reporting systems are crucial due to the critical nature of tasks performed under stress and time constraints. Ali et al. (2022) highlighted the significance of user-centered design in emergency applications, emphasizing intuitive navigation and prompt information retrieval during emergencies. Their research underscores the necessity for rigorous usability evaluations to improve the performance and user acceptance of emergency reporting systems [13].

2.5 Application of the PACMAD Model in Web and Mobile Applications

The PACMAD model has proven effective across various domains of web and mobile applications for assessing usability and user experience. Rey (2023) applied the PACMAD model to evaluate the usability of criminal records tracking applications, while Tan (2023) utilized it in assessing automotive on-demand services. These studies underscore the versatility of the PACMAD model in evaluating user interaction and satisfaction across different contexts [14, 15].

3 Methodology

3.1 Research Design

A descriptive research design was employed to comprehensively evaluate the usability of the WeBarangay web application using the PACMAD framework. This design allowed for a detailed exploration and assessment of the application's usability dimensions, including effectiveness, efficiency, satisfaction, learnability, memorability, errors, and cognitive load. It was chosen based on its suitability for identifying strengths, weaknesses, and areas for improvement in WeBarangay to enhance local governance and service delivery.

3.2 Participants

Selection Criteria

Participants included residents of Barangay N.S. Amoranto, who actively used the WeBarangay web application, barangay officials responsible for administering WeBarangay, and technical administrators involved in its maintenance and development. Criteria for selection prioritized individuals who had used WeBarangay for at least three months to ensure familiarity with its functionalities and those who voluntarily participated in the study.

Justification

Residents provided insights into user experience and satisfaction, while barangay officials and technical administrators offered perspectives on administrative functionalities and technical aspects, contributing to a comprehensive evaluation of WeBarangay's usability.

Table 1 Frequency and the Percentage of the respondents

Respondents	Sample	
	Frequency	Percentage (%)
Residents	35	70
Barangay officials	10	20
Technical administrators	5	10
Total	50	100

Sample Distribution

The study included 50 respondents, distributed as follows: 70% residents (35 individuals), 20% barangay officials (10 individuals), and 10% technical administrators (5 individuals), as shown in Table 1. Residents provided insights into user experience, barangay officials contributed administrative perspectives, and technical administrators offered insights on technical aspects. This distribution ensured a comprehensive evaluation of the WeBarangay application's usability and functionality across different stakeholder groups.

3.3 Data Collection Methods

Instrumentation

Data collection methods included:

- Surveys: A structured questionnaire based on the PACMAD framework was administered to assess user perceptions of effectiveness, efficiency, satisfaction, learnability, memorability, errors, and cognitive load. Likert scales, as shown in Table 2, ranging from 1 (Not Acceptable) to 5 (Highly Acceptable), were used to measure participants' responses across these dimensions. The intervals are based on sources such as Dawes (2008) [16] and Pimentel (2010) [17]. The study sample consisted of 50 respondents, and the questionnaire was distributed online using Google Forms.

Table 2 Likert scale weight

Numerical rating	Numerical scale	Descriptive rating
5	4.21–5.00	Highly acceptable
4	3.41–4.20	Very acceptable
3	2.61–3.40	Acceptable
2	1.81–2.60	Fairly acceptable
1	1.00–1.80	Not acceptable

Table 3 Reliability statistics

Cronbach's alpha	Cronbach's alpha based on standardized items	N of items
0.901	0.902	50

- Interviews: Semi-structured interviews with barangay officials and technical administrators were conducted to gather qualitative insights on usability issues, administrative functionalities, and technical challenges.
- Usability Testing: Task-based usability testing sessions were conducted to observe and measure user interactions with WeBarangay, focusing on task completion rates, time taken, and user errors.

The reliability of the questionnaire was assessed using Cronbach's alpha test, which indicated the measuring instrument's high reliability with a value of 0.901, as shown in Table 3. Statistical analysis was conducted using SPSS software (Taherdoost 2016) [18]. The raw Cronbach's alpha of 0.901 and the standardized alpha of 0.902 indicate strong reliability and coherence in measuring the intended construct. Researchers can have confidence in the consistency of responses, affirming the effectiveness of the questionnaire in assessing the targeted aspects of the study.

Rationale

Surveys with Likert scales provided quantitative data on user satisfaction and perceptions of usability dimensions, enabling statistical analysis. Interviews complemented these findings with qualitative insights, while usability testing validated the results through direct observation of user interactions.

3.4 Procedure

Implementation steps

1. Participant Recruitment: Participants were identified and recruited based on the selection criteria outlined in Sect. 3.2.
2. Data Collection: Surveys were administered to residents, and interviews were conducted with barangay officials and technical administrators. Usability testing sessions were scheduled and conducted.
3. Data Analysis: Survey responses were compiled and analyzed using descriptive and inferential statistics. Interview transcripts were coded and analyzed for themes and patterns. Usability testing results were analyzed to identify usability issues and performance metrics.
4. Reporting: Findings were summarized, conclusions drawn, and recommendations formulated based on the analysis.

3.5 Variables and Measures

Independent and dependent variables

1. Independent Variables: Usability dimensions (effectiveness, efficiency, satisfaction, learnability, memorability, errors, cognitive load) assessed using the PACMAD framework.
2. Dependent Variables: User satisfaction, task completion rates, time taken to complete tasks, frequency and severity of errors, and user feedback on usability issues.

3.6 Data Analysis

Data Processing and Analysis

Quantitative data from surveys were analyzed using descriptive statistics (e.g., means, frequencies) to summarize usability scores across dimensions. Inferential statistics (e.g., correlation analysis, regression) were employed to explore relationships between usability dimensions and user satisfaction. Qualitative data from interviews and usability testing underwent thematic analysis to identify key themes, usability challenges, and improvement suggestions.

4 Results and Discussion

4.1 Comprehensive Usability Evaluation Using PACMAD Framework

Effectiveness: Analysis of task completion rates and success metrics.

The effectiveness evaluation of WeBarangay, as shown in Table 4, indicated high user satisfaction across key metrics. Users found the number of steps to complete tasks (EFE1) and task completion within defined time limits (EFE2) very acceptable, scoring 3.98 and 4.00, respectively. They also reported minimal errors during tasks (EFE3), with a score of 4.18. The overall mean score of 4.05 suggested consistent and effective performance, highlighting WeBarangay's reliability and efficiency in supporting barangay-related activities.

Efficiency: Time taken to complete tasks and resource utilization.

Table 5 summarizes the efficiency evaluation of WeBarangay. Users found the time taken to complete tasks (EFF1) very acceptable, scoring 4.12. Navigation efficiency on each page (EFF2) received a score of 4.14, indicating high satisfaction. Users also reported minimal interactions (EFF3) with a score of 4.16. Overall, WeBarangay

Table 4 Effectiveness evaluation results

ID	Effectiveness categories	Weighted mean	Standard deviation	Verbal interpretation
EFE1	How many steps do you take to finish a given task?	3.98	0.58	Very acceptable
EFE2	How many tasks have been solved in a predefined time?	4.00	0.48	Very acceptable
EFE3	How many errors have occurred during the task?	4.18	0.32	Very acceptable
	Overall mean	4.05	0.46	Very acceptable

achieved a mean score of 4.14, highlighting consistent efficiency and user satisfaction across all metrics. These findings underscored the application's effectiveness in facilitating smooth and efficient task completion within the barangay, positively impacting user experience.

Satisfaction: User feedback and Likert scale responses.

Table 6 presents the satisfaction evaluation results for WeBarangay. Users rated their overall experience using the application (SA1) with a weighted mean of 4.38, indicating a highly acceptable level of satisfaction. The standard deviation of 0.42 suggests consistent positive feedback among users. Additionally, satisfaction with the application itself (SA2) received a score of 4.24, further confirming high levels of user satisfaction with WeBarangay. The overall mean score across both categories was 4.31, reinforcing the application's effectiveness in meeting user expectations and delivering a positive user experience. These results underscore WeBarangay's success in satisfying users' needs and preferences, contributing positively to its usability and user engagement.

Table 5 Efficiency evaluation results

ID	Efficiency categories	Weighted mean	Standard deviation	Verbal interpretation
EFF1	How much time is taken to complete a given task?	4.12	0.42	Very acceptable
EFF2	How long taken on each page included in the task	4.14	0.46	Very acceptable
EFF3	How many clicks (touch) to solve the task?	4.16	0.38	Very acceptable
	Overall mean	4.14	0.42	Very acceptable

Table 6 Satisfaction evaluation results

ID	Efficiency categories	Weighted mean	Standard deviation	Verbal interpretation
SA1	How do the users feel when using the application?	4.38	0.42	Highly acceptable
SA2	Are the users satisfied with the application?	4.24	0.46	Highly acceptable
	Overall mean	4.31	0.44	Highly acceptable

Learnability: Ease of understanding and navigating WeBarangay.

Table 7 presents the learnability evaluation results for WeBarangay. Users rated the ease of learning the application (LEA1) with a weighted mean of 4.28, indicating a highly acceptable level of ease, with a standard deviation of 0.62, suggesting some variability in responses. The time taken for users to learn the application (LEA2) received a score of 3.96, which is considered very acceptable, with a standard deviation of 0.44, indicating relatively consistent feedback. The overall mean score for learnability was 4.12, reflecting a very acceptable level of user agreement on the application's learnability. These results highlight that WeBarangay is generally easy to learn and quick for users to become proficient, enhancing its overall usability.

Memorability: User recall of features and functions.

Table 8 presents the memorability evaluation results for the WeBarangay application. Users rated two key aspects of memorability: firstly, the ease of reusing the application without assistance (MEM1), which received a weighted mean score of 4.16 (SD = 0.52), indicating a very acceptable level. This suggests that users found it straightforward to remember how to use the application independently. Secondly, the number of tasks achieved during the first use (MEM2) was rated at 3.98 (SD = 0.46), which is also considered very acceptable, reflecting users' ability to accomplish tasks effectively upon initial interaction. The overall mean score for memorability was 4.07 (SD = 0.49), reinforcing the application's strong performance in this aspect. These results indicate that WeBarangay maintains a high level of memorability, with users finding it easy to remember and reuse, which is crucial for enhancing user experience and satisfaction over time.

Table 7 Learnability evaluation results

ID	Learnability categories	Weighted mean	Standard deviation	Verbal interpretation
LEA1	Is the application easy to learn?	4.28	0.62	Highly acceptable
LEA2	How much time does it take for the user to learn?	3.96	0.44	Very acceptable
	Overall mean	4.12	0.53	Very acceptable

Table 8 Memorability evaluation results

ID	Memorability categories	Weighted mean	Standard deviation	Verbal interpretation
MEM1	How easy is it to reuse the application without help?	4.16	0.52	Very acceptable
MEM2	How many tasks were achieved during the first use?	3.98	0.46	Very acceptable
	Overall mean	4.07	0.49	Very Acceptable

Error Frequency and Severity: Identification and categorization of user errors.

Table 9 presents the error evaluation results for the WeBarangay application. Users rated three key aspects related to error handling. Firstly, the application's ability to assist users in recovering from errors (ERR1) received a high score of 4.18 (SD = 0.38), indicating effective support. Secondly, the clarity of error messages in guiding users on recovery (ERR2) scored 4.10 (SD = 0.34), showing clear and helpful communication. Thirdly, the application's responsiveness in resolving errors quickly (ERR3) was rated 4.20 (SD = 0.42), indicating prompt problem-solving. Overall, with an average score of 4.16 (SD = 0.38), WeBarangay demonstrates strong error management capabilities, ensuring a smooth user experience by effectively handling errors and providing clear guidance.

Cognitive Load: User perceptions of mental effort and workload.

Table 10 presents the cognitive load evaluation results for WeBarangay. Users rated the achievement of tasks within the given time (COG1) with a weighted mean of 4.40, indicating a highly acceptable level of performance with a standard deviation of 0.42. The impact of errors on user workload (COG2) scored 4.64, which is also highly acceptable, with a standard deviation of 0.44, suggesting minimal disruption due to errors. The increase in mental effort during tasks (COG3) was

Table 9 Errors evaluation results

ID	Errors evaluation categories	Weighted mean	Standard deviation	Verbal interpretation
ERR1	Is the application help the user to recover from errors?	4.18	0.38	Very acceptable
ERR2	Is the error messages inform the user how to recover?	4.10	0.34	Very acceptable
ERR3	Is the app responding easily and quickly to recover errors?	4.20	0.42	Very acceptable
	Overall mean	4.16	0.38	Very acceptable

Table 10 Cognitive load evaluation results

ID	Cognitive load categories	Weighted mean	Standard deviation	Verbal interpretation
COG1	Has the user achieved all the given tasks within the given time?	4.40	0.42	Highly acceptable
COG2	Is the user workload affected by an error during the task?	4.64	0.44	Highly acceptable
COG3	Is the user's mental effort increased during the task?	4.48	0.48	Highly acceptable
COG4	Is the user performance stable during the tasks?	4.18	0.44	Very acceptable
COG5	How much time is taken by the user to respond?	4.28	0.46	Highly acceptable
	Overall mean	4.40	0.45	Highly acceptable

rated at 4.48, reflecting a highly acceptable level, with a standard deviation of 0.48. User performance stability (COG4) received a score of 4.18, which is considered very acceptable, with a standard deviation of 0.44. Lastly, the time taken by users to respond (COG5) was rated at 4.28, highly acceptable, with a standard deviation of 0.46. The overall mean score for cognitive load was 4.40, indicating a highly acceptable level of cognitive demand. These results demonstrated that users found WeBarangay to be cognitively manageable, with stable performance and minimal mental strain during tasks.

4.2 *Impact on Communication Channels*

Improvement in Communication: Insights from Resident and Official Feedback

Residents and barangay officials reported significant improvements in communication through WeBarangay. The satisfaction evaluation results (Table 6) showed high ratings, with users feeling positive about the application (SA1) scoring 4.38 and general satisfaction (SA2) scoring 4.24. These high scores indicated that WeBarangay effectively facilitated clear and efficient communication, enhancing information dissemination and community engagement.

Effectiveness of Announcements: Assessment of Message Reach and Clarity

The effectiveness of announcements on WeBarangay was evidenced by user feedback on task completion and error frequency. The effectiveness evaluation (Table 4) showed very acceptable scores for task steps (EFE1) at 3.98 and task completion time (EFE2) at 4.00. Minimal errors during tasks (EFE3) scored 4.18, indicating that announcements and instructions were clear and easy to follow. The overall mean score of 4.05 confirmed the application's capability to communicate important messages effectively.

User Engagement: Analysis of Interaction Levels and Response Rates

User engagement with WeBarangay was analyzed through efficiency and cognitive load evaluations. Efficiency results (Table 5) showed timely task completion (EFF1) at 4.12, efficient page navigation (EFF2) at 4.14, and minimal interactions (EFF3) at 4.16. The overall efficiency score of 4.14 indicated active user engagement without unnecessary delays. The cognitive load evaluation (Table 10) showed users successfully achieved tasks within the given time (COG1) at 4.40, with minimal impact from errors (COG2) scoring 4.64. The overall cognitive load score of 4.40 indicated manageable mental effort, contributing to sustained engagement and efficient interaction with the application.

4.3 Strengths and Weaknesses in Service Delivery

Service Efficiency

Residents and barangay officials found WeBarangay's service delivery to be efficient, with a high overall mean efficiency score of 4.14 (Table 5). Tasks were completed quickly, with minimal interactions required, reflecting prompt and responsive service.

Administrative Workflow

Barangay officials reported that WeBarangay is well integrated into their daily operations. The satisfaction evaluation showed a high overall mean score of 4.31 (Table 6), indicating positive feedback on usability and learnability. The application was easy to learn (4.28) and quick to adopt, enhancing administrative efficiency (Table 7).

Technical Performance

Technical administrators noted that WeBarangay generally performed well, but some stability issues were identified. While error recovery was rated highly (4.20, Table 9), cognitive load evaluation indicated room for improvement in performance stability (4.18, Table 10). Ongoing technical refinements are needed to enhance robustness and reliability.

4.4 User Feedback and Areas for Improvement

Common User Issues: Themes from Qualitative Data and Survey Responses

User feedback revealed common issues such as difficulties with initial setup, occasional glitches, connectivity problems, and slow response times. Despite these challenges, overall satisfaction remained high.

Functionality Enhancements: Recommendations for Feature Improvements

Users recommended adding real-time notifications for announcements and a robust search function for past documents. Officials suggested integrating detailed analytics and reporting tools to monitor community engagement and service usage better.

Design Modifications: Suggestions for Interface Adjustments

Users proposed a cleaner, more intuitive layout with customizable dashboards, larger buttons, and a simplified color scheme. Additionally, they recommended adding a tutorial or help section to assist new users in navigating the application.

4.5 Actionable Recommendations for Optimization

Prioritized Recommendations

Key recommendations based on evaluation findings and stakeholder feedback include enhancing real-time notifications for timely updates, addressing connectivity issues, and optimizing response times. Implementing advanced search functionalities and redesigning the interface for better usability were also highlighted. Additionally, creating comprehensive user tutorials and manuals to aid navigation and conducting regular performance audits were suggested.

Implementation Strategies

Strategies for integrating improvements involve incremental feature development with rigorous testing before deployment. Training sessions, user manuals, and ongoing user feedback mechanisms will be crucial for adoption and adaptation. Continuous monitoring of performance metrics and proactive maintenance will ensure sustained functionality and user satisfaction.

Long-Term Usability Goals

Long-term goals focus on enhancing user experience through iterative interface enhancements and feature updates. Integration of analytics for informed decision-making and robust security measures to protect user data are priorities. Future-proofing the platform with scalable infrastructure and community-centric features aims to foster engagement and support effective governance practices.

5 Conclusion

In conclusion, the evaluation of WeBarangay through the PACMAD framework has provided valuable insights into its impact on local governance and community engagement. The study aimed to assess the effectiveness of WeBarangay in facilitating communication between barangay officials and residents, enhancing transparency, and improving overall governance processes.

Based on the findings, WeBarangay has shown considerable promise in addressing the communication gaps within barangay administrations. The application's user-friendly interface and accessibility have contributed significantly to its adoption among residents and officials alike. Through surveys and user feedback, it became evident that WeBarangay has facilitated quicker response times to community concerns, increased participation in local initiatives, and fostered a sense of inclusivity among residents.

Furthermore, the PACMAD evaluation framework enabled a comprehensive analysis of WeBarangay's functionalities, focusing on its Performance, Accessibility, Compatibility, Maintainability, Adaptability, and Dependability aspects. While the application scored well across these dimensions, areas for improvement were identified, particularly in enhancing notification systems and addressing occasional technical glitches reported by users.

Recommendations for future enhancements include refining the notification features to ensure timely updates, conducting regular maintenance to improve system reliability, and exploring additional features that could further empower residents to participate actively in barangay affairs. These improvements are crucial for sustaining WeBarangay's effectiveness and relevance in the evolving landscape of digital governance tools.

References

1. Rey WP, Aumentado JDP, Delos Rey JNC, Dipad MLD (2024) WeBarangay: a local E-government services web application for N.S. Amoranto in Quezon City, Philippines. In: Proceedings of the 2024 10th international conference on computing and data engineering (ICCDE '24). Association for Computing Machinery, New York, pp 80–88. https://doi.org/10.1145/3641181.3641199
2. Harrison R, Flood D, Duce D (2013) Usability of mobile applications: literature review and rationale for a new usability model. J Interact Sci 1(1):1. https://doi.org/10.1186/2194-0827-1-1
3. Nielsen J (2012) Usability 101: introduction to usability. Nielsen Norman Group logo-Nielsen Norman Group. Retrieved from: https://www.nngroup.com/articles/usability-101-introduction-to-usability/
4. Goel S, Nagpal R, Mehrotra D (2018) Mobile applications usability parameters: taking an insight view. https://doi.org/10.1007/978-981-10-3932-4_4
5. Alturki R, Gay V (2017) Usability testing of fitness mobile application: methodology and quantitative results. Comput Sci Inf Technol 97–114. https://doi.org/10.5121/csit.2017.71108

6. Ali W, Riaz O, Mumtaz S, Khan AR, Saba T, Bahaj SA (2022) Mobile application usability evaluation: a study based on demography. IEEE Access 10:41512–41524. https://doi.org/10.1109/ACCESS.2022.3166893
7. Kortum P, Sorber M (2015) Measuring the usability of mobile applications for phones and tablets. Int J Human-Comput Interact 31(8):518–529. https://doi.org/10.1080/10447318.2015.1064658
8. Moumane K, Idri A, Abran A (2016) Usability evaluation of mobile applications using ISO 9241 and ISO 25062 standards. Springerplus 5(1):548. https://doi.org/10.1186/s40064-016-2171-z
9. Kumar BA, Goundar MS (2019) Usability heuristics for mobile learning applications. Educ Inf Technol 24(2):1819–1833. https://doi.org/10.1007/s10639-019-09860-z
10. Saleh A, Isamil RB, Fabil NB (2015) Extension of Pacmad model for usability evaluation metrics using goal question metrics (Gqm) approach. J Theor Appl Inform Technol 79(1)
11. Az-zahra HM, Fauzi N, Kharisma AP (2019) Evaluating E-marketplace mobile application based on people at the center of mobile application development (PACMAD) usability model. In: 2019 international conference on sustainable information engineering and technology (SIET). IEEE, pp 72–77. https://doi.org/10.1109/SIET48054.2019.8986067
12. Parsazadeh N, Ali R, Rezaei M, Tehrani SZ (2018) The construction and validation of a usability evaluation survey for mobile learning environments. Stud Educ Eval 58:97–111. https://doi.org/10.1016/j.stueduc.2018.06.002
13. Almasi S, Mehrabi N, Asadi F, Afzali M (2022) Usability of emergency department information system based on users' viewpoint; a cross-sectional study. Arch Acad Emerg Med 10(1):e71. https://doi.org/10.22037/aaem.v10i1.1635
14. Rey WP (2023) Assessing MABIS Mobile app based on people at the center of mobile application development (PACMAD) usability model: empirical investigation. In: 2023 13th international conference on software technology and engineering (ICSTE), Osaka, Japan, pp 37–43. https://doi.org/10.1109/ICSTE61649.2023.00014
15. Rey WP, Del Rosario EJ, Lasquety MK, Tan KAD (2023) X-Mech: an on-demand vehicle express repair service mobile application. In: 2023 13th international conference on software technology and engineering (ICSTE), Osaka, Japan, pp 93–99. https://doi.org/10.1109/ICSTE61649.2023.00023
16. Dawes J (2008) Do data characteristics change according to the number of scale points used? An experiment using 5-point, 7-point, and 10-point scales. Int J Mark Res 50(1):61–104. https://doi.org/10.1177/147078530805000106
17. Pimentel JL (2010) A note on the usage of Likert Scaling for research data analysis. USM R&D J 18(2):109–112. https://www.researchgate.net/profile/JonaldPimentel/publication/331231816_A_note_on_the_usage_of_Likert_Scaling_for_research_data_analysis/links/5f82ff6592851c14bcbe847d/A-note-on-theusage-of-Likert-Scaling-for-research-dataanalysis.pdf
18. Taherdoost H (2016) Validity and reliability of the research instrument; how to test the validation of a questionnaire/survey in a research. Int J Acad Res Manage (IJARM) 5(3):28–36. https://doi.org/10.2139/ssrn.3205040

DOA Estimation Based on Robust Generalized Chinese Remainder Theorem for Monostatic MIMO Radar

Mei Zhang, Yongbo Zhao, and Chenghu Cao

Abstract Compared with the traditional array, the sparse array can offer a larger array aperture with the same number of physical sensors. In this paper, an efficient method to estimate direction of arrivals (DOAs) of two targets based on robust generalized Chinese remainder theorem (RGCRT) for monostatic MIMO radar in sparse array is proposed. Firstly, we employ the all-phase time-shifting phase difference correcting method (AP-TSPDC) to determine the wrapped phases of two targets in the noise environments. And then the improved RGCRT is utilized for the phase unwrapping, enabling accurate DOA estimation for both targets. Besides, the proposed algorithm displays high applicability due to no strict constraint on the array aperture. The future research will extend the proposed algorithm to bistatic radar models. The proposed algorithm is able to achieve high DOA estimation precision with low computational complexity. Computer simulation results confirm the effectiveness and stability of the proposed algorithm.

Keywords Monostatic MIMO radar · DOA estimation · All-phase time-shifting phase difference correcting method · Wrapped phase · Robust generalized Chinese remainder theorem

1 Introduction

As a significant electromagnetic sensor, the radar has tremendous applications in various areas [1, 2], such as precision guidance, airborne early warning, weather remote sensing, road traffic and so on. Multiple-input multiple-output (MIMO) radar, an advanced system integrating techniques from communication and digital array

---

M. Zhang · Y. Zhao (✉)
National Key Laboratory of Radar Signal Processing, Xidian University, Xi'an, China
e-mail: ybzhao@xidian.edu.cn

C. Cao
School of Electronic Engineering, Xi'an University of Posts and Communications, Xi'an, China

© The Author(s), under exclusive license to Springer Nature Switzerland AG 2025
S. Femmam and P. Lorenz (eds.), *Recent Advances in Communication Networks and Embedded Systems*, Lecture Notes on Data Engineering and Communications Technologies 256, https://doi.org/10.1007/978-3-031-92449-1_7

signal processing, offers enhanced capabilities over the traditional radar by independently controlling transmitting antenna waveforms [3]. DOA estimation, as an important topic in MIMO radar signal processing, has been widely studied in recent years.

MIMO radar with sparse array structure has a larger array aperture than which with traditional array, which can not only reduce the problem of mutual coupling between array elements, but also improve the resolution of DOA to make the target detection performance better. However, the array spacing does not conform to the Nyquist sampling theorem, i.e., the array spacing is larger than half wavelength, thereby leading to the failure of angle estimation. A kind of coprime array structure is proposed [4, 5], which need to satisfy the special array structure. In [6], a unitary estimation of signal parameter via rotation invariance techniques (ESPRIT) algorithm based on coprime array MIMO radar was proposed, which utilizes the coprime property to deal with the problem of DOA ambiguity. Moreover, another method based on wavelet operator is respectively applied to super-resolution searching algorithms, such as multiple signal classification (MUSIC) and ESPRIT [7]. In [8], a model of sparse array is designed to improve the angular measurement performance. The methods mentioned above have certain constraints and limitations on the array structure.

The main problem of DOA estimation in sparse array is phase ambiguity. To solve the problem, some algorithms have been proposed. Previous studies [9–11] have proposed various robust Chinese remainder theorems for resolving single-objective ambiguities without closed-form solutions. Moreover, a closed-form Chinese remainder theorem was introduced in [12], capable of reconstructing folded integers into a single integer. As mentioned in [13], the most important problem to deal with multi-targets phase unwrapping is to establish one-to-one correspondence between ambiguous phases and targets. In [14], a DOA estimation based on CRT for multi-targets is proposed. the limitation is that the method can only handle the ambiguity of one target. Hence, the doppler information is required to assist the matching of ambiguous phases and targets.

Discrete Fourier transform (DFT) can be applied to spectrum analysis of signals. In order to suppress the spectrum leakage, an all-phase DFT (APDFT) is proposed to obtain the true phase [15, 16]. A time-shift phase difference correction method based on APDFT (AP-TSPDC) is proposed in [17], which is convenient to obtain accurate estimation of the initial phases.

In this paper, we perform an improved RGCRT algorithm to estimate the DOAs of two targets for monostatic radar. The subspace theory is not employed so that the proposed algorithm displays excellent performance when handling adjacent targets. It is worth mentioning that the algorithm possesses good applicability due to less stringent constraints and limitations on the array structure.

2 Problem Statement

2.1 The Signal Model of Monostatic Radar

The geometry of the monostatic MIMO radar is shown in Fig. 1. Consider the radar system equipped with M_s transmit antennas and M_r receive antennas, and M_s transmit antenna transmits orthogonal waveform $\mathbf{S} = \left[\mathbf{s}_1, \mathbf{s}_2, \ldots, \mathbf{s}_{M_s}\right]^{\mathrm{T}} \in \mathbb{C}^{M_s \times U}$ where U is the number of snapshots of the received data. Assuming that there are L far-field targets impinging on the receive arrays from directions θ_l, $l = 1, \ldots, L$. The received signal can be expressed as

$$X = \sum_{l=1}^{L} \beta_l \mathbf{a}_s(\theta_l)\mathbf{a}_r^{\mathrm{T}}(\theta_l)\mathbf{S} + \mathbf{W} \tag{1}$$

where β_l denotes the reflection coefficient of the lth target, $\mathbf{a}_s(\theta_l)$ is the transmit steering vector of the lth target, $\mathbf{a}_r(\theta_l)$ is the receive steering vectors of the lth target, $\mathbf{a}_s(\theta_l) = \left[1, e^{j2\pi \sin\theta_l d_{t1}/\lambda}, \ldots, e^{j2\pi \sin\theta_l (d_{t1}+d_{t2}+\cdots+d_{tM_s})/\lambda}\right]^{\mathrm{T}}$, λ is the wavelength, $\mathbf{a}_r(\theta_l) = \left[1, e^{j2\pi \sin\theta_l d_{r1}/\lambda}, \ldots, e^{j2\pi \sin\theta_l (d_{r1}+d_{r2}+\cdots+d_{rM_r})/\lambda}\right]^{\mathrm{T}}$, the inter-sensor spacing of transmit array is larger than half of the wavelength, and the inter-sensor spacing of receive array is larger than half of the wavelength, $d > \lambda/2$, $\mathbf{W}$ is a complex Gaussian white noise with zero mean, the notation $(\cdot)^{\mathrm{T}}$ denotes the transpose operation.

After time-domain matched filtering, the output signal is obtained as follows

$$\overline{\mathbf{X}} = \sum_{l=1}^{L} \beta_l \boldsymbol{a}_s(\theta_l)\boldsymbol{a}_r^{T}(\theta_l) + \overline{\mathbf{W}} \tag{2}$$

where $\overline{\mathbf{W}} = \mathbf{W}\mathbf{S}^{\mathrm{H}}/U$ represents the noise matrix after matched filtering, the notation $(\cdot)^{\mathrm{H}}$ denotes the Hermitian transpose. Assume that there are Q pulses of $\overline{\mathbf{X}}$. Vectorize $\overline{\mathbf{X}}$ to obtain the matrix $\mathbf{Y}$

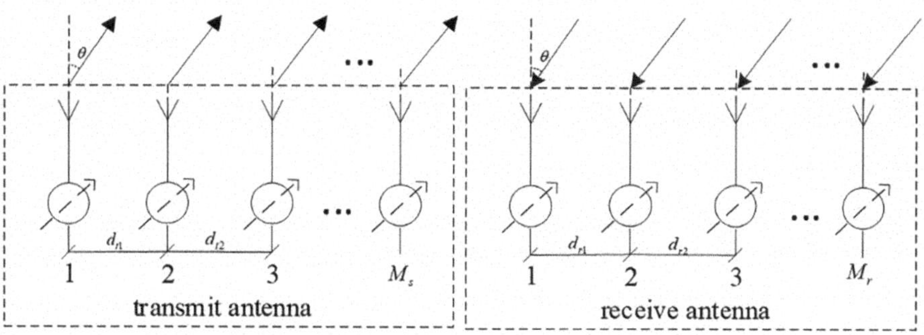

Fig. 1 The geometry of the monostatic radar

$$\mathbf{Y} = \left[y_1, y_2, \cdots, y_Q\right] = \mathbf{AC} + \overline{\mathbf{W}} \tag{3}$$

where $\mathbf{y}_q$ is the output of vectorizing the qth pulse, $\mathbf{A} = [\mathbf{a}(\theta_1), \mathbf{a}(\theta_2), \cdots, \mathbf{a}(\theta_L)] \in \mathbb{C}^{M_s M_r \times L}$ represents the receive-transmit steering matrix, $\mathbf{a}(\theta_l)$ is the receive-transmit steering vector of the lth target, $\mathbf{a}(\theta_l) = \mathbf{a}_s(\theta_l) \otimes \mathbf{a}_s(\theta_l) \in \mathbb{C}^{M_s M_r \times 1}$, $\mathbf{C} = \left[\mathbf{c}_1, \mathbf{c}_2, \cdots, \mathbf{c}_Q\right] \in \mathbb{C}^{L \times Q}$ is the matrix of reflection coefficients, $\mathbf{c}_q$ is the vector of L targets reflection coefficients in the qth pulse, $\otimes$ denotes Kronecker product.

2.2 *The Problem of DOA Estimation Utilizing Super Resolution Algorithm*

The model of monostatic MIMO radar in sparse array is described above. In order to explain the equivalent relation of array structure, suppose that $M_s = 2$, $M_r = 3$, the equivalent virtual antenna is shown in Fig. 2.

When the inter-sensor spacing exceeds half the wavelength, i.e., $d > \lambda/2$, conventional super-resolution techniques are impractical for directly resolving the true DOAs from two sources from the multiple spectral peaks distributed uniformly. The limitation arises due to the large inter-sensor spacing and the periodicity of the exponential function of steering matrix. Hence, there is a need to find a DOA estimation algorithm suitable for universal sparse array, thereby improving the applicability and precision of DOA estimation while ensuring low computation.

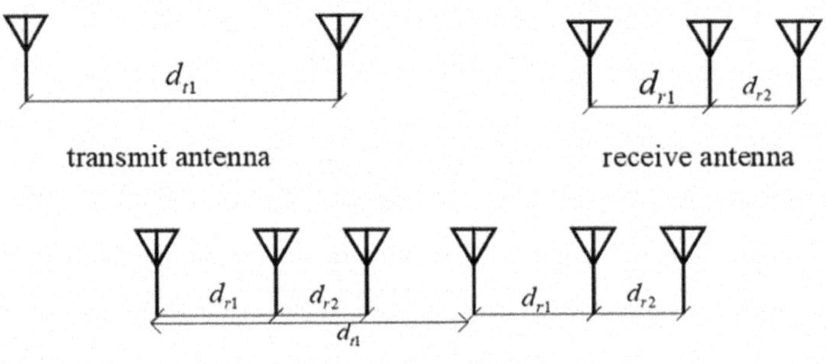

Fig. 2 The equivalent virtual antenna

3 DOA Estimation Based in Improved RGCRT

3.1 The Correspondence Between DOA Estimation and CRT

As shown in Fig. 2, suppose that the spacing between adjacent sensors in array is not equal, and the spacing between p th and $(p+1)$ th element is d_p, $1 \le p \le N-1$, where $d_p > \lambda/2$, N is the number of equivalent virtual antenna. Suppose that two far-field targets impinging on the receive arrays from directions θ_l, $l = 1, 2$. The phase difference of lth target between the p th and $(p+1)$ th element is written as

$$\Delta\varphi_{l,p} = \frac{2\pi d_p \sin\theta_l}{\lambda} \tag{4}$$

Owing to the phase ambiguity, the relationship between the real phase and apparent phase is

$$\Delta\varphi_{l,p} = 2\pi \times n_{l,p} + \Delta\widetilde{\varphi}_{l,p}, n_{l,p} \in \mathbb{Z}^+ \tag{5}$$

where $n_{l,p}$ is the folding integer, $\Delta\widetilde{\varphi}_{l,p}$ is the apparent phase. Equation (6) can be obtained by combining Eqs. (4) with (5).

$$\frac{2\pi d_p sin\theta_l}{\lambda} = 2\pi \times n_{l,p} + \Delta\tilde{\varphi}_{l,p}, n_{l,p} \in \mathbb{Z}^+ \tag{6}$$

The following parameters can be defined based on the array structure

$$\begin{cases} d_0 = C \times M \times \prod_{p=1}^{P} m_p \\ d_p = \frac{C}{m_p} \times \prod_{p=1}^{P} m_p \end{cases} \tag{7}$$

Both sides of Eq. (7) are multiplied by the parameter $d_0/(2\pi d_p)$, Eq. (8) can be written as

$$\frac{d_0 sin\theta_l}{\lambda} = \frac{d_0}{d_p} \times n_{l,p} + \frac{d_0 \Delta\tilde{\varphi}_{l,p}}{2\pi d_p}, n_{l,p} \in \mathbb{Z}^+ \tag{8}$$

Equation (8) can be simplified as

$$\begin{cases} N_1 = n_{1,p} T_p + \hat{r}_{1,p} \\ N_2 = n_{2,p} T_p + \hat{r}_{2,p} \end{cases}, 1 \le p \le N-1 \tag{9}$$

where $N_l = \frac{d_0 \sin\theta_l}{\lambda}$, $\hat{r}_{l,p} = \frac{d_0 \Delta\widetilde{\varphi}_{l,p}}{2\pi d_p}$, $T_p = \frac{d_0}{d_p}$, $\hat{r}_{l,p} = r_{l,p} + \Delta r_{l,p}$, $l = 1, 2, 1 \le p \le N-1$, $\Delta r_{l,p}$ denotes the remainder errors, $r_{l,p}$ denotes error-free remainders.

3.2 The Correspondence Between DOA Estimation and CRT

Suppose that there is a greatest common divisor G between the modular values in Eq. (9), Eq. (9) can be written as

$$\begin{cases} N_1^0 = n_{1,p} m_p + q_{1,p} \\ N_2^0 = n_{2,p} m_p + q_{2,p} \end{cases}, 1 \le p \le N-1 \tag{10}$$

where $N_l^0 = \left\lfloor \frac{N_l}{G} \right\rfloor$, $m_p = \left\lfloor \frac{T_p}{G} \right\rfloor$, $q_{l,p} = \left\lfloor \frac{\hat{r}_{1,p}}{G} \right\rfloor$, $l = 1, 2$. Moreover, the common remainders are defined as

$$\hat{r}^c_{l,p} = \hat{r}_{l,p} \bmod G \tag{11}$$

The $2P$ common remainders are sorted in the increasing order in the following

$$\hat{r}^c_{\sigma(1)} \le \hat{r}^c_{\sigma(2)} \le \cdots \le \hat{r}^c_{\sigma(2P)} \tag{12}$$

where σ denotes the permutation of the set $\{1, 2, \ldots, 2P\}$. The distance between the adjacent common remainders is defined as

$$D_p = \begin{cases} \hat{r}^c_{\sigma(p+1)} - \hat{r}^c_{\sigma(p)}, p = 1, 2, \cdots, 2P-1 \\ \hat{r}^c_{\sigma(1)} - \hat{r}^c_{\sigma(2P)} + G, p = 2P \end{cases} \tag{13}$$

According to [18], the lemma is given as follows.

Lemma 1 Define $\tau = \max \left|\Delta r_{l,p}\right|, l = 1, 2, p = 1, 2, \cdots, P$, if $\tau < G/8$, there exists one and only one subscript $p_0 \in \{1, 2, \cdots, P\}$ satisfies.

$$D_{p_0} + D_{p_0+P} > \frac{G}{2} \tag{14}$$

Two sets are defined below, which satisfy $w_P - w_1 \le 2\tau$, $v_P - v_1 \le 2\tau$.

$$\Psi_1 = \{w_1, w_2, \ldots, w_P\} = \left\{\hat{r}^c_{\sigma(p_0+1)}, \hat{r}^c_{\sigma(p_0+2)}, \cdots, \hat{r}^c_{\sigma(p_0+P)}\right\}$$

$$\Psi_2 = \{v_1, v_2, \ldots, v_P\} = \begin{cases} \left\{\hat{r}^c_{\sigma(1)}, \hat{r}^c_{\sigma(2)}, \cdots, \hat{r}^c_{\sigma(P)}\right\}, p_0 = P \\ \left\{\hat{r}^c_{\sigma(p_0+1+P)} - G, \cdots, \hat{r}^c_{\sigma(2P)} - G, \hat{r}^c_{\sigma(1)}, \cdots, \hat{r}^c_{\sigma(p_0)}\right\} p_0 \ne P \end{cases} \tag{15}$$

Two common remainders can be calculated according two sets defined above

$$\overline{w}_1 = \frac{w_1' + w_2' + \cdots + w_P'}{P}, \overline{w}_2 = \frac{v_1 + v_2 + \cdots + v_P}{P} \tag{16}$$

where $w_p' = \begin{cases} w_p, w_P - v_1 \leq G/2 \\ w_p - G, w_P - v_1 > G/2 \end{cases}$. Then the estimates of integers in Eq. (11) is represented as

$$\hat{q}_{l,p} = \left[\frac{\hat{r}_{l,p} - \overline{w}_t}{G}\right], l = 1,\ 2; p = 1,\ 2, \ldots, P \tag{17}$$

where $t = \begin{cases} 1, d_G\left(\hat{r}^c_{l,p}, w_{p_1}\right) = 0 \\ 2, d_G\left(\hat{r}^c_{l,p}, w_{p_2}\right) = 0 \end{cases}$, $d_G(x, y)$ is circular distance, which is defined as

$$d_G(x, y) = x - y - \left[\frac{x - y}{G}\right]G \tag{18}$$

Explanation about how to reconstruct the real-value set $\left\{N_1^0, N_2^0\right\}$ from the remainder set $\left\{\hat{q}_{l,p}, l = 1,\ 2, p = 1,\ 2, \ldots, P\right\}$ is at below. In the terms of the basic operation properties of modular, $\left\{N_1^0, N_2^0\right\}$, satisfies the following relations

$$\begin{cases} N_1^0 + N_2^0 = \langle N_1^0 + N_2^0 \rangle_{m_p} = \langle \hat{q}_{1,p} + \hat{q}_{2,p} \rangle_{m_p} mod m_p \\ N_1^0 N_2^0 = \langle N_1^0 N_2^0 \rangle_{m_p} = \langle \hat{q}_{1,p} \hat{q}_{2,p} \rangle_{m_p} mod m_p \end{cases} \tag{19}$$

The following equations are defined as

$$\begin{cases} \mathcal{M} = \prod_{p=1}^{P} m_p \\ \Gamma_p = \frac{}{m_p} \end{cases} \tag{20}$$

As described in [19], the set $\left\{N_1^0, N_2^0\right\}$ is obtained by

$$\begin{cases} N_1^0 = \frac{\zeta_1 - \sqrt{\zeta_1^2 - 4\zeta_2}}{2} \\ N_2^0 = \frac{\zeta_1 + \sqrt{\zeta_1^2 - 4\zeta_2}}{2} \end{cases} \tag{21}$$

where $\begin{cases} \zeta_1 = N_1^0 + N_2^0 = \langle \sum_{p=1}^{P} \langle q_{1,p} + q_{2,p} \rangle_{m_p} \Gamma_p \overline{\Gamma}_p \rangle \mathcal{M} \\ \zeta_2 = N_1^0 N_2^0 = \langle \sum_{p=1}^{P} \langle (q_{1,p} - s) + (q_{2,p} - s) \rangle_{m_p} \Gamma_p \overline{\Gamma}_p \rangle \mathcal{M} \end{cases}$, $s = \max\left\{0, \left\lceil \zeta_1 - 2\sqrt{M} \right\rceil\right\}$. Due to the uncertainty of the correspondence between $\left\{N_1^0, N_2^0\right\}$ and $\{\overline{w}_1, \overline{w}_2\}$, the minimum mean square error criterion is employed to estimate the two common remainders, which are denoted as follows

$$\begin{cases} \hat{r}_1^c = arg \min\limits_{x \in [c_1, c_2]} \sum_{i=1}^{I} \|x - \hat{r}^c_{1,p_i}\|^2 \\ \hat{r}_2^c = arg \min\limits_{x \in [c_3, c_4]} \sum_{i=1}^{I} \|x - \hat{r}^c_{2,p_i}\|^2 \end{cases} \tag{22}$$

where $c_1 = \min\left\{\hat{r}^c_{1,p_1}, \cdots, \hat{r}^c_{1,p_l}\right\}$, $c_2 = \max\left\{\hat{r}^c_{1,p_1}, \cdots, \hat{r}^c_{1,p_l}\right\}$. $c_3 = \min\left\{\hat{r}^c_{2,p_1}, \cdots, \hat{r}^c_{2,p_l}\right\}$, $c_4 = \max\left\{\hat{r}^c_{2,p_1}, \cdots, \hat{r}^c_{2,p_l}\right\}$. Till now, the real phases of two targets are estimated as

$$\begin{cases} \widehat{N}_1 = N_1^0 G + \hat{r}^c_1 \\ \widehat{N}_2 = N_2^0 G + \hat{r}^c_2 \end{cases} \tag{23}$$

where $\widehat{N}_1$, $\widehat{N}_2$ denote the estimation of true phases of two targets, the DOA of the two targets can be obtained from the following equation

$$\hat{\theta}_l = \arcsin\frac{\widehat{N}_1\lambda}{d_0}, l = 1,2 \tag{24}$$

From the elaboration above, it can be seen that the improved RGCRT algorithm will implement the phase unwrapping with remainder errors. The subsequent section showcases how to extract the apparent phase.

3.3 All-Phase Time-Shifting Phase Difference Correcting

Consider a complex exponential sequence with length $2N_{DFT} - 1$ and amplitude A, $\{x(n) = A \cdot \exp(j(w_0 n + \theta_0)), n \in [-N_{DFT} + 1, N_{DFT} - 1]\}$, and the sequence with time-delay $n_0\{x_1(n) = A \cdot \exp(j(w_0(n - n_0) + \theta_0)), n \in [-N_{DFT} + 1, N_{DFT} - 1]\}$, . The spectral analysis and correction of two sequences are performed below. Firstly, $x(n)$ is transferred to the frequency domain utilizing the APFFT method [15], which is written as

$$F(k) = A\exp\big(j\theta\big) W_n^2 (k\Delta w - w_0\big) \tag{25}$$

Due to the linear time invariant properties of the APFFT, $x_1(n)$ can be transferred to

$$F_1(k) = A\exp\big(j(\theta - w_0 n_0) W_n^2 (k\Delta w - w_0\big) \tag{26}$$

Thus, the phase difference of two sequences located on the main spectral lines is

$$\Delta\varphi = w_0 n_0 = \varphi_1\big(s^*\big) - \varphi_2\big(s^*\big) + \frac{2s^*\pi n_0}{N_{DFT}} \tag{27}$$

where $\varphi_1(s^*)$ and $\varphi_2(s^*)$ represent the phase of two sequences at the main spectral lines, $\frac{2s^*\pi n_0}{N_{DFT}}$ is the phase compensation value caused by the time delay. Hence, the formula of the amplitude correction can be obtained as follows

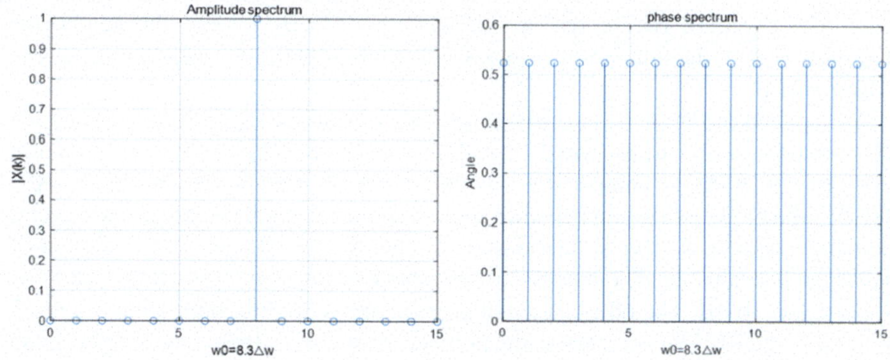

Fig. 3 Spectrum of AP-TSPDC

$$A = \frac{|F(s^*)|}{W_n^2(dw)} \tag{28}$$

The performance of the algorithm is shown in Fig. 3, the time-delay $n_0 = N_{DFT}$. As shown in Fig. 3, it can be concluded that as an optimized algorithm, AP-TSPDC possesses a superior spectral correction accuracy, which can efficaciously mitigate the impact of spectral interference and spectral leakage. That is to say, the amplitude spectrum demonstrates outstanding performance in the noise environment.

4 Numerical Experiments

In this section, several simulations are presented to verify the DOA estimation performance of the proposed algorithm. To illustrate the performance of the algorithm, the proposed algorithm is compared with the DOA estimation based on joint singular value decomposition [20], the ESPRIT for coprime array [21], the RGCRT for sparse array [22], the improved root-MUSIC for co-prime array [23]. The wavelength is $\lambda = 0.3$m, the root mean square error (RMSE) is used to measure the performance of algorithms, which is defined as

$$\text{RMSE} = \sqrt{\frac{1}{J}\frac{1}{L}\sum_{j=1}^{J}\sum_{l=1}^{L}\left(\widehat{\theta}_{jl} - \theta_l\right)^2} \tag{29}$$

where $\widehat{\theta}_{jl}(1 \le j \le J, 1 \le l \le L)$ denotes the DOA estimate of lth target in jth Mont Carlo trial, θ_l is the real DOA of lth target. In the first simulation, the performance of DOA estimation versus different SNR is presented. The number of transmit antennas is $M_s = 4$ with adjacent element spacing $d_1 = 2.2m, d_2 = 1.4m, d_3 = 1m$. The number of receive antennas is $M_r = 3$ with uniform adjacent element spacing $d =$

0.6*m*. The DOAs of two targets are set as $\theta_1 = 20°$, $\theta_2 = 40°$, the number of snapshots is 200, the number of Monte Carlo trials is 500, the variation of SNR is 0 dB to 25 dB.

As depicted in Fig. 4, it is clearly observable that all the algorithms are valid for DOA estimation, and the proposed algorithm demonstrates slightly higher estimation precision of DOAs in comparison with the other algorithms. Owing to the AP-TSPDC, the phase extraction becomes more accurate as the SNR increases, thereby reducing the spectral leakage. In the next simulation, the influence of the number of snapshots is evaluated. The number of snapshots varies from 10 to 1000, the SNR is located as 10 dB. The other parameters are same as the first simulation. Figure 5 shows that the RMSE decreases as the number of snapshots increases, and the value of RMSE tend to be stable when the number of snapshots reaches 200.

Fig. 4 RMSE versus SNR

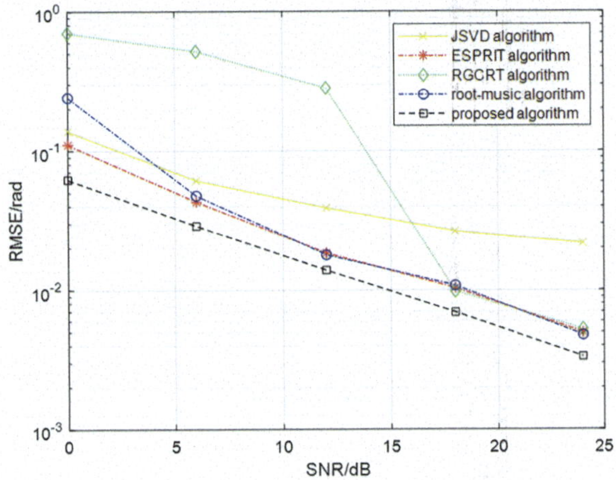

Fig. 5 RMSE versus the number of snapshots

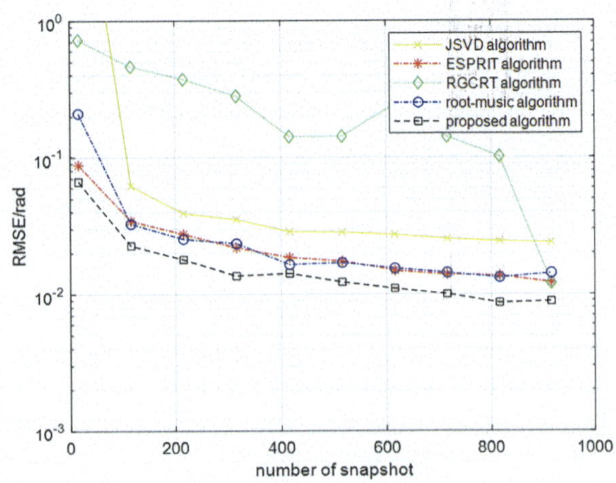

In the following simulation, the computational complexity of all algorithms is indicated. The number of snapshots is set as 200, the SNR is 10 dB, the number of Monte Carlo trials varies from 5 to 100, the other parameters are same as the first simulation. In Fig. 6, it can be seen that the proposed algorithm takes relatively less time because other algorithms require the searching operation.

In this simulation, the size of array aperture is considered. The number of snapshots is 1000, the adjacent element spacing of transmit antennas is set as $d_1 = 2.2m, d_2 = 1.4m, d_3 = 1m$ and $d_1 = 1.4m, d_2 = 1m$ respectively, the SNR varies from 0 to 40 dB, the other parameters are same as the first simulations. As Fig. 7 shows, the angle measurement error will diminish with the increase of array aperture to a certain extent.

Fig. 6 Running time versus the number of Monte Carlo trials

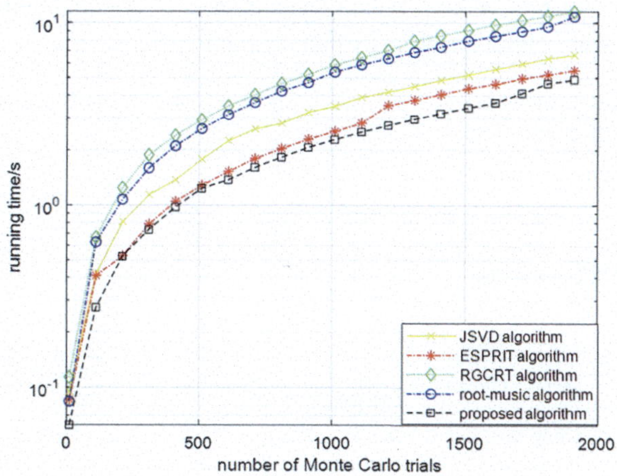

Fig. 7 RMSE versus SNR

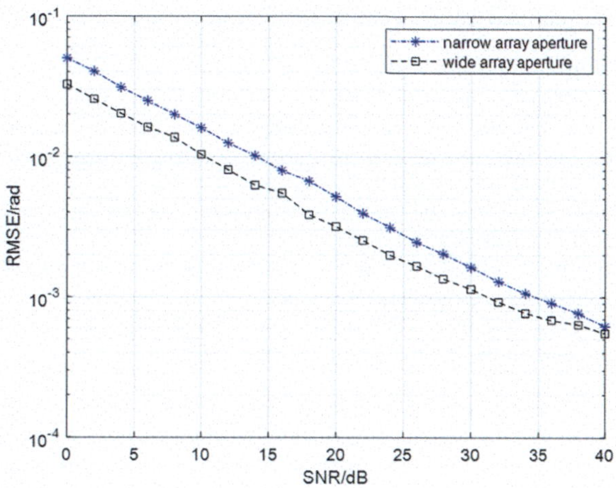

5 Conclusions

In this paper, the DOA estimation of two targets for sparse array monostatic MIMO radar is presented. Firstly, the algorithm utilizes the AP-TSPDC algorithm for spectrum analysis and correction to obtain the wrapped phases of two targets. Subsequently, the improved RGCRT is employed to resolve the phase ambiguity of the sparse array, thereby determining the DOA estimates of two targets. The theoretical analysis and computer simulation indicate that the method enhances DOA estimation precision of two targets including close targets while maintaining low computational complexity.

Since the proposed algorithm imposes no significant constraints on the array structure, it shows high applicability. Hence, the algorithm focused on the model of monostatic MIMO radar can be applied to the angle measurement of bistatic MIMO radar, which requires further research.

Acknowledgements This work was supported by the National Natural Science Foundation of China under Grant 62271379.

References

1. Lu TY, Wang L, Shang T, Liu YP (2018) The guidance strategy for maneuvering target with phased array radar seeker. In: Proceedings of the 2018 IEEE CSAA guidance, navigation and control conference. IEEE
2. Chandrasekar V, Chen HN, Philips B (2018) DFW urban radar network observations of floods, tornadoes and hail storms. In: Proceedings of the 2018 IEEE radar conference. IEEE, pp 765–770
3. Zhao YB, Liu HW (2018) Overview on MIMO radar. J Data Acquisit Proc 33(03):389–399
4. Vaidyanathan PP, Pal P (2010) Sparse sensing with coprime arrays. In: Proceedings of 2010 conference record of the forty fourth asilomar conference on signals, systems and computers. IEEE, pp 1405–1409
5. Zhou C, Gu Y, Fan X, Shi ZG, Mao GP, Zhang YD (2018) Directions-of-arrivals estimations for coprime array via virtual array interpolation. IEEE Trans Signal Proc 66(22):5956–5971
6. Li J, Jiang D, Zhang X (2017) DOA estimation based on combined unitary ESPRIT for coprime MIMO radar. IEEE Commun Lett 21(1):96–99
7. Mao XJ, Pan HH (2013) An improved DOA estimation algorithm based on wavelet operator. J Commun 8(12):839–844
8. Qin G, Amin M, Zhang Y (2019) DOA estimation exploiting sparse array motion. IEEE Trans Signal Process 67(11):3013–3027
9. Li G, Xu J, Peng YN, Xia XG (2007) An efficient implementation of a robust phase-unwrapping algorithms. IEEE Signal Proc Lett 14(6):393–396
10. Li XW, Xia XG (2008) A fast robust Chinese remainder theorem based phase unwrapping algorithm. IEEE Signal Proc Lett 15(10):665–668
11. Li XW, Liang H, Xia XG (2009) A robust Chinese remainder theorem with its applications in frequency estimation from under sampled waveforms. IEEE Trans Signal Proc 57(11):4314–4322
12. Wang W, Xia XG (2010) A closed-form robust Chinese remainder theorem and its performance analysis. IEEE Trans Signal Proc 58(11):5655–5666

13. Xia XG (1999) On estimation of multiple frequencies in under sampled complex valued waveforms. IEEE Trans Signal Proc 47(12):3417–3419
14. Cao CH, Zhao YB, Pang XJ, Chen S, Hu YL (2020) Novel DOAs estimation method based on Doppler aided Chinese remainder theorem with all phase DFT for multiple targets in sparse array. IET Radar Sonar Navig 14(10):1512–1520
15. Huang XD, Wang ZH (2007) Principle of all-phase DFT restraining spectral leakage and the application in correcting spectrum. J Tianjin Univ 40(7):882–886
16. Wang ZH, Huang XD (2009) Principle of phase measurement and its application based on all-phase spectral analysis. J Data Acquis Proc 24(6):777–782
17. Li XF, Li L, Kou K, Wu TF (2016) Analysis and improvement of time-shift phase difference spectral correction based on all-phase FFT. J Tianjin Univ 49(12):1290–1295
18. Li XP, Xia XG, Wang WJ, Wang W (2016) A robust generalized Chinese remainder theorem for two integers. IEEE Trans Inf Theory 62(12):7491–7504
19. Wang W, Li X, Xia XG, Wang WJ (2015) The largest dynamic range of a generalized Chinese remainder theorem for two integers. IEEE Signal Proc Lett 22(2):254–258
20. Yang XP, Wu XC, Li Sarkar TK (2018) A fast and robust DOA estimation method based on JSVD for co-prime array. IEEE Access 6:41697–41705
21. Sun FG, Gao B, Chen LZ, Lan P (2016) A low-complexity ESPRIT-based DOA estimation method for co-prime linear arrays. Sensors 16(9):1367
22. Li XP, Cao YH, Yao BB, Liu F (2018) Robust generalized Chinese-remainder-theorem-based DOA estimation for a coprime array. IEEE Access 6:2169–3536
23. Zhang D, Zhang YS, Zheng GM, Feng CQ, Tang J (2017) Improved DOA estimation algorithm for co-prime linear arrays using root-MUSIC algorithm. Electron Letter 53(18):1277–1279

Emerging Network Technology and Information Security

Leveraging Graph Neural Network for Attack Detection Systems in Heterogeneous Network Data

Samia Saidane, Francesco Telch, Kussai Shahin, and Fabrizio Granelli

Abstract In this paper, we introduce Multi-Layer-GraphSAGE algorithm, a modified version of the GraphSAGE model tailored specifically for detecting attacks in heterogeneous datasets. With the proliferation of cyber threats across diverse network environments, there is a pressing need for robust and adaptable detection mechanisms. We conduct extensive experiments using multiple datasets. By prioritizing graph topologies over features, ML-GraphSAGE overcomes frequent obstacles in intrusion detection systems (IDS), such as difficulty recognizing minority classes in imbalanced datasets. The experimental results show that our proposed technique performs well across a variety of datasets. It achieves an average accuracy of 94.33% on the CIC-DDoS2019 dataset with 13 classes; 94.68% on the InSDN dataset with 8 classes; and 95.59% on the Edge-IIoT dataset with 15 classes.

Keywords Intrusion detection · Graph neural network · NetFlow data · Graph representation learning · Edge classification

S. Saidane (✉) · F. Granelli
DISI, University of Trento, Trento, Italy
e-mail: samia.saidane@unitn.it

F. Granelli
e-mail: fabrizio.granelli@unitn.it

F. Telch · K. Shahin
Trentino Digitale Spa, Trento, Italy
e-mail: francesco.telch@tndigit.it

K. Shahin
e-mail: kussai.shahin@tndigit.it

© The Author(s), under exclusive license to Springer Nature Switzerland AG 2025
S. Femmam and P. Lorenz (eds.), *Recent Advances in Communication Networks and Embedded Systems*, Lecture Notes on Data Engineering and Communications Technologies 256, https://doi.org/10.1007/978-3-031-92449-1_8

1 Introduction

Cyberattacks are not only increasing in frequency but also growing in complexity. The challenge for businesses is to effectively mitigate these attacks and prevent information loss. Detecting attacks before execution has become a crucial task. Initially, Network Intrusion Detection Systems (NIDSs) were introduced as a promising approach to monitor network traffic and flows, offering two types: signature-based and anomaly based. Signature-based NIDS [1] can identify known attacks based on predefined rules in its database. However, it has limitations, such as the inability to recognize unknown attacks or different variants of known attacks that do not match any signature in the rules. Research integrates advanced ML and DL into NIDS to address evolving threats. However, commercial adoption lags due to AI solutions' unreliable performance in real-world settings, exacerbated by poor model generalization to diverse or unseen network scenarios [2]. Furthermore, Conventional NIDS architectures analyze network flows individually, neglecting interdependencies. A general model adaptable to diverse datasets, capturing intricate network relationships, is crucial for detecting complex attacks spanning multiple activities. One promising avenue for enhancing NIDS capabilities is the adoption of Graph Neural Networks (GNNs). GNNs are specifically designed to operate on graph-structured data, making them well-suited for modelling the complex interactions inherent in network traffic [3]. This paper emphasizes addressing data heterogeneity, advocating for adaptable general models and GNN adoption over conventional deep learning. This paper focuses on the following contributions: *-ML-GraphSAGE for Detecting Attacks in Heterogeneous Datasets*: We introduce ML-GraphSAGE, a novel approach specifically designed to detect attacks in datasets containing diverse types of network traffic. *-Handling Imbalanced Datasets*: We address the challenge of imbalanced datasets common in intrusion detection systems (IDS), enhancing classification effectiveness, particularly for minority classes. *-Reducing False Positives and False Negatives*: ML-GraphSAGE aims to mitigate the high false positive rates (FPR) and false negative rates (FNR) typically associated with conventional IDS approaches. The rest of the paper is organized as follows: Sect. 2 presents a review of previous relevant works. The proposed ML-GraphSAGE model is described in Sect. 3. The proposed methodology for intrusion detection systems based on ML-GraphSAGE is detailed in Sect. 4. Experimental evaluation results are presented in Sects. 5, and 6 provides conclusions and future works.

2 Related Works

2.1 Conventional Approaches for Network Intrusion Detection

Reference [4] compared seven ML algorithms across NIDS datasets, finding unsupervised methods like IF and one-class SVM showed better generalization than supervised approaches like SGD-one-class SVM. Reference [5] hybridized C5 and SVM classifiers for a HIDS, detecting attack behavior using labels and anomalies. Reference [6] developed a bi-directional LSTM network with sequence and feature layers to detect attacks across various scales. Historical data aids in learning attack patterns, enhancing identification across different scopes. Suzen et al. [7] introduced a hybrid DBN model using Contrastive Divergence (CD) for hidden layer updates, achieving high accuracy in detecting industrial Internet attacks. Liang et al. [8] devised an industrial NIDS algorithm using a multifeatured data clustering optimization model. The authors in [9] proposed a data-driven intrusion detection method combining time-domain and frequency-domain analyses. In [10], FELIDS, a federated learning-based IDS for agricultural-IoT, utilized deep learning (DNNs, CNNs, RNNs) achieving high accuracy on CSE-CIC-IDS2018 and InSDN datasets.

2.2 Graph Based Approach for Network Intrusion Detection

Lo et al. [11] proposed E-GraphSAGE for IoT NIDS, extending GraphSAGE. It achieved perfect results in binary tasks but encountered challenges in multi-class scenarios due to possible redundancy in neighborhood aggregation, affecting identification accuracy. In [12], Alert-GCN utilized GCNs to predict attacks by linking related alerts. While achieving good results on DARPA99, the custom similarity method for edge definition may have caused overfitting in the graph neural network due to high self-similarity among neighboring nodes. Caville et al. [13] introduced Anomal-E, an intrusion and anomaly detection method based on E-GraphSAGE, achieving over 97.5% accuracy and at least 92% F1-Score on NF-UNSWNB15-v2 and NF-CSE-CIC-IDS2018-v2 datasets. In [14], authors proposed a GNN model for end-to-end botnet detection, emphasizing network topology over node and edge features. They integrated genuine background traffic into botnet datasets for binary classification to identify botnets [15]. Approaches in [12, 13] used standard graph embedding by converting network flows into graphs with IP and port pairs, employing methods like DeepWalk. However, they rely on transudative embedding, limiting generalization to unseen nodes.

3 Multi-layer GraphSAGE Model

In this section, we introduce Multi-Layer GraphSAGE, a variation of the classic GraphSAGE model, tailored to address specific challenges in network analysis, particularly in the context of intrusion detection systems (NIDS). Classic GraphSAGE [16, 17] focuses on NIDS edge classification, integrating edge features in embeddings to differentiate benign and attack flows. In contrast, ML-GraphSAGE incorporates custom message passing, multi-layer stacking, an MLP predictor, and class weights calculation to improve predictive capabilities and handle imbalanced data.

3.1 Edge Embedding

In the Multi-Layer GraphSAGE (ML-GraphSAGE) model, edge embeddings are computed to capture the representation of each edge in the graph. The edge embedding process involves concatenating the embeddings of the source and destination nodes connected by the edge. Mathematically, the edge embedding $\mathbf{Z}_{uv}^k$ for each edge (u, v) at layer k can be represented as the concatenation of the node embeddings $\mathbf{Z}_u^k$ and $\mathbf{Z}_v^k$:

$$\mathbf{Z}_{uv}^k = \text{CONCAT}(\mathbf{Z}_u^k, \mathbf{Z}_v^k) \quad \forall\ (u, v) \in \varepsilon \tag{1}$$

Here, $\mathbf{Z}_{uv}^k$ represents the embedding for edge (u, v) at layer k, $\mathbf{Z}_u^k$ and $\mathbf{Z}_v^k$ represent the embeddings of nodes u and v respectively, and CONCAT denotes the concatenation operation.

3.2 Sampling and Aggregation

In ML-GraphSAGE, sampling and aggregation are performed to gather information from neighboring nodes and edges, which is crucial for generating node and edge embeddings.

Sampling: The sampling process involves selecting a set of neighboring nodes and edges around each node. This can be done using various sampling strategies such as uniform sampling or random walk sampling.

Aggregation: In ML-GraphSAGE, a custom aggregation function is employed, which considers both node and edge features. Mathematically, the aggregation function $h_k^N(v)$ for node v at layer k can be represented as:

$$h_k^N(v) = \text{AGG}_k\left(\text{ne}_{uv}^{(k-1)}, \forall u \in N(v), uv \in E_0\right) \tag{2}$$

Here, $ne_{uv}^{(k-1)}$ represents the features of edge uv from the sampled neighborhood of node v at layer $k-1$, $N(v)$ represents the sampled edges about node v, and E_0 denotes the set of all edges in the graph. AGG_k denotes the aggregation function.

4 NIDS Based on ML GraphSAGE

In this paper, we propose a Graph Neural Network (GNN) model using Multi-Layer GraphSAGE for intrusion detection. The datasets used in our study encompass a wide range of intrusion scenarios. The intrusion detection methodology (Fig. 1) begins with data preprocessing and graph generation, followed by training a model using ML-GraphSAGE. The trained model produces a feature matrix for rich node representations, and Gtest is then used for classifying data flows and attacks.

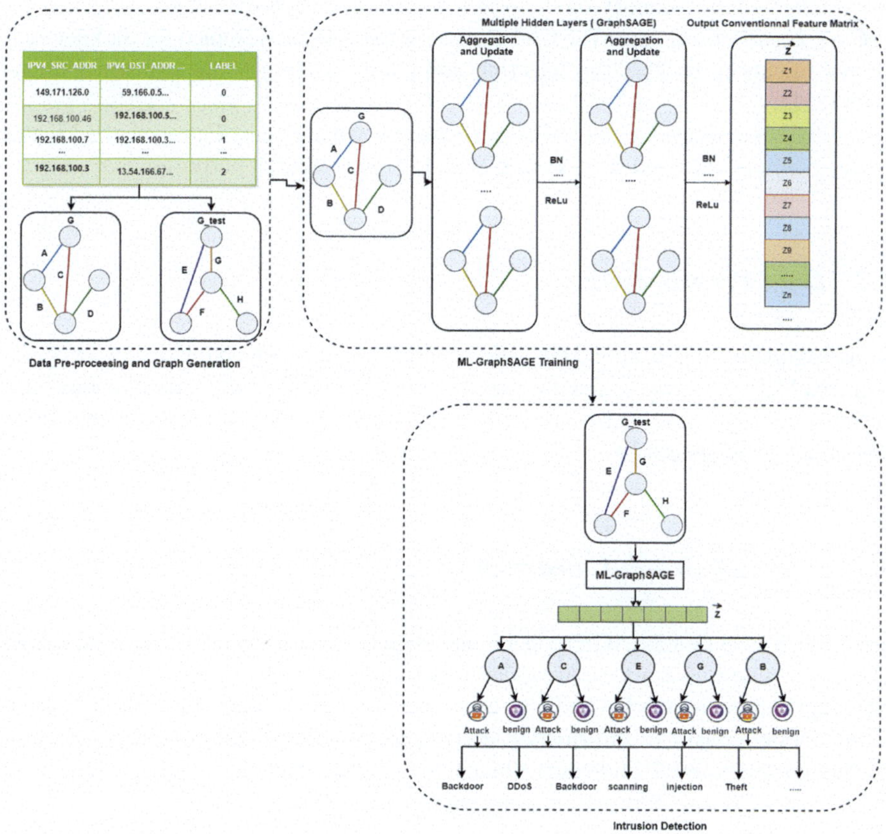

Fig. 1 Proposed ML-GraphSAGE for intrusion detection

4.1 Datasets

NF-UNSW-NB15: The 2015 UNSW-NB15 dataset collected by the Cyber Range Lab of ACCS includes 87.35% benign and 12.65% attack flows, totalling 2,540,044 flows [18].
CSE-CIC-IDS2018: The 2018 CSE CIC dataset includes 83.07% benign and 16.93% attack flows, simulating a network with five organizational departments and a server room [19].
CIC DDoS2019: The CIC-DDoS2019 dataset simulates real-world data with benign and modern DDoS attack instances, using CICFlowMeter-V3 to generate labelled flows with timestamps, IP addresses, ports, protocols, and attack vectors [20].
InSDN: The InSDN dataset, created by University College Dublin in September 2020, addresses issues like redundancy and is the first publicly dataset generated using the SDN environment [21].
NF-UQ-NIDS: A comprehensive dataset, merging four datasets. It includes an additional label feature, identifying the original dataset of each flow [18].
CIC-darknet2020 Dataset: The CIC-Darknet2020 dataset, designed for darknet traffic classification, focuses on detecting and characterizing VPN and Tor applications [22].
Edge-IIoTset dataset: The Edge-IIoTset dataset is generated by Ferrag et al. [23] and can be publicly accessed from Ferrag et al. [23] for research purposes.

4.2 Data Preprocessing

To prepare for ML-GraphSAGE, the dataset converts IP addresses to integers, removes Null/Nan values, and encodes categorical features. Numerical features are normalized using a standard scalar for computational efficiency and independence from fixed IP addresses.

4.3 Network Graph Construction

The Multi-Layer GraphSAGE model is designed to process network data represented as a graph, where each node corresponds to a unique (IP address, Port) tuple and each edge represents a network flow between source and destination nodes. Node features in this model consist of information derived from the application layer source and destination ports and IP addresses. On the other hand, edge features are derived from the remaining traffic information associated with each network flow.

4.4 ML-GraphSAGE Training

For Multi-Layer GraphSAGE training, network data is graphically represented with nodes as (IP, Port) tuples and edges indicating flows. Node features (e.g., app layer details of source/dest IPs and ports) and edge features (e.g., remaining traffic specifics) are extracted and stored. Initialization includes node features, edge features, and a binary training mask for edges.

4.5 Edge Classification

After the model parameters have been trained, the ML-GraphSAGE model is ready for evaluation through the classification of unseen test samples. Test flow records are converted into graphs, passed through the trained ML-GraphSAGE layers to calculate edge embeddings, and then processed through a softmax layer to compute class probabilities. The model classifies edges in the graph, predicting whether each network flow is normal or anomalous based on learned features. Each edge is assigned a probability score indicating its likelihood of belonging to a specific class.

5 Results and Discussion

In this section, we present experiment results of intrusion detection with Multi-layer GraphSAGE using Python 3 and Jupyter notebooks. Seven diverse datasets (Table 1) validate our system across different environments. For each dataset, we split the data into a training set comprising 70% of the examples and a testing set comprising 30% of the examples.

Table 1 Overview of the datasets used in our experiments

Dataset	No. examples	Normal traffic samples (%)	No. classes	No. features
NF-UNSW-NB15	98494	96.83	10	43
CIC-DarkNet2020	158616	82.82	4	85
CSE-CIC-IDS	601901	70.61	15	77
NF-UQ-NIDS	1019914	76.78	21	77
CIC-DDoS2019	440000	4.30	13	74
InSDN	343889	19.11	8	84
Edge-IIoTset	2219201	81.59	15	63

Table 2 Classification report for NF-UNSW-NB15

Label	Accuracy (%)	Precision (%)	Recall (%)	F1-score (%)
Analysis	94.96	9.95	37.74	15.72
Backdoor	94.91	12.36	33.71	18.09
Benign	88.58	98.40	77.00	86.40
DoS	95.34	16.13	5.25	7.92
Exploits	89.99	74.95	57.41	65.02
Fuzzers	92.20	92.82	53.00	67.47
Generic	94.67	35.63	25.72	29.87
Reconnaissance	96.88	98.24	70.20	81.88
Shellcode	85.11	7.56	95.05	14.01
Worms	92.81	1.54	93.55	3.03
Weighted average	92.34	83.38	61.74	70.50

5.1 *NF-UNSW-NB15 Dataset Results*

The evaluation metrics for the intrusion detection model trained on the NF-UNSW-NB15 dataset, exhibit a mixed performance as shown in Table 2 and in the Confusion Matrix in Fig. 2. The class "Benign" achieves the highest precision of 98.40%, indicating a strong ability to correctly classify normal network traffic. Additionally, "Reconnaissance" demonstrates high precision (98.24%) and recall (70.20%), with an F1-score of 81.88%, showcasing its effectiveness in identifying reconnaissance activities. The weighted average accuracy of 92.34% reflects the model's general effectiveness in classifying instances from the NF-UNSW-NB15 dataset.

Figure 3 shows the True Positive Rate (TPR) and False Positive Rate (FPR) for various classes. The x-axis lists the classes: Analysis, Backdoor, Benign, DoS, Exploits, Fuzzers, Generic, Reconnaissance, Shellcode, and Worms. The y-axis represents the rates in percentage. Blue bars indicate TPR, with the highest values observed for Shellcode (97.62%) and Worms (93.55%). Red bars indicate FPR, with the highest values for Shellcode (5.03%) and Worms (7.18%).

5.2 *NF-UQ-NIDS Dataset*

In the evaluation of an intrusion detection model based on a Multi-layer GraphSAGE architecture using the NF-UQ-NIDS dataset, several classes demonstrate noteworthy performance metrics as shown in Table 3 and in the Confusion Matrix (CM) in Fig. 4. "Benign" showcases high precision (99.50%) and an F1-score of 59.54%, indicate its effectiveness in correctly identifying normal network and IoT traffic. Additionally, "Reconnaissance" displays strong precision (76.52%) and decent recall (52.16%), leading to an F1-score of 61.32%, highlighting its capability in detecting reconnais-

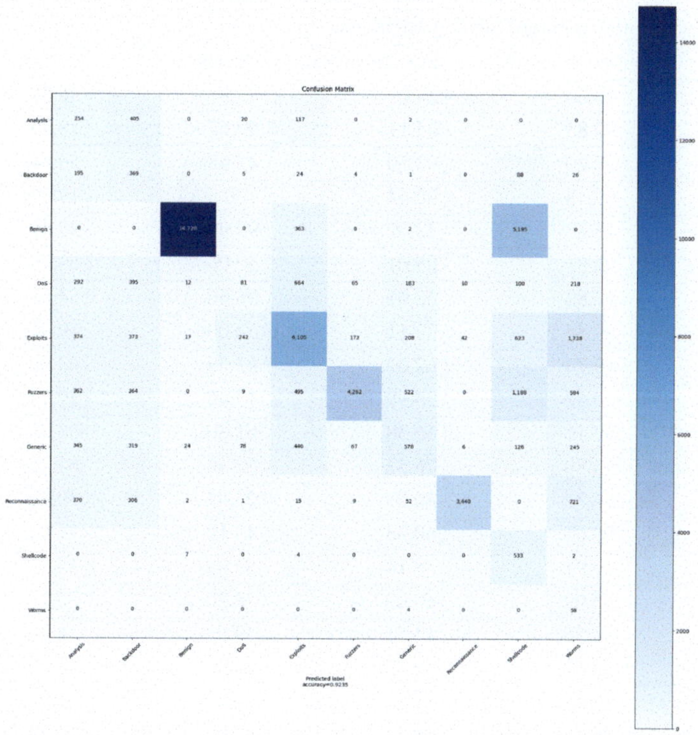

Fig. 2 CM with NF-UNSW-NB15

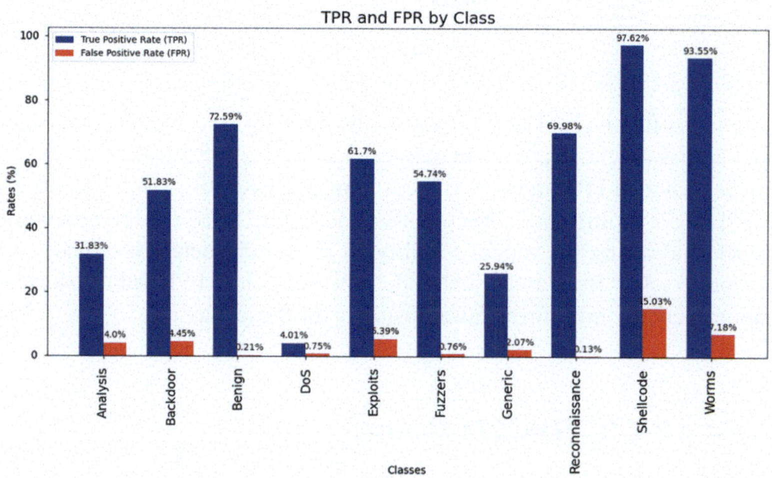

Fig. 3 TPR versus FPR by class

Table 3 Classification report of NF-UQ-NIDS data

Label	Accuracy (%)	Precision (%)	Recall (%)	F1-score (%)
Analysis	90.60	24.0	73.81	36.79
Backdoor	99.83	65.11	99.57	78.73
Benign	70.20	99.50	42.48	59.54
Bot	99.48	90.44	100	17.25
Brute Force	75.88	23.02	99.99	37.42
DDoS	94.81	85.61	50.06	63.18
DoS	93.82	38.01	96.46	54.53
Exploits	99.44	32.11	52.00	39.41
Fuzzers	99.74	62.02	63.33	62.68
Generic	99.52	36.99	13.46	19.74
Infilteration	97.16	56.38	61.80	58.96
Reconnaissance	95.80	76.52	52.16	61.32
Shellcode	96.79	30.03	93.41	50.87
Theft	99.05	60.33	43.19	11.95
Worms	99.91	8.11	59.68	14.37
Injection	95.09	60.00	28.20	38.17
MITM	99.56	11.51	49.81	18.70
Password	92.77	53.53	32.93	40.78
Ransomware	99.98	33.53	100	50.22
Scanning	96.10	8.24	66.05	14.64
XSS	94.83	30.00	47.48	36.52
Average	98.53	40.66	60.88	50.34

sance activities effectively. The accuracy analysis reveals the overall performance of the model across all classes, with an average accuracy of 98.53%.

From the provided TPR and FPR values in Fig. 5, classes like "Bot," "Brute Force," "DDoS," "DoS," "Infiltration," "Reconnaissance," and "XSS" exhibit relatively high TPR values, indicating the model's ability to accurately detect instances of these classes. Conversely, "Bot" and "Shellcode" showcase low FPR values suggest that the model effectively minimizes false positives for these classes.

5.3 *NF-CSE-CIC-IDS2018 Dataset*

Intrusion detection using the NF-CSE-CIC-IDS2018 dataset, encompassing 15 classes, reveals varying performance across different classes as illustrated in Table 4 and in the confusion matrix in Fig. 6. Notably, "Bot" demonstrates exceptional precision (99.33%), recall (100.00%), and F1-score (99.67%), indicating the model's

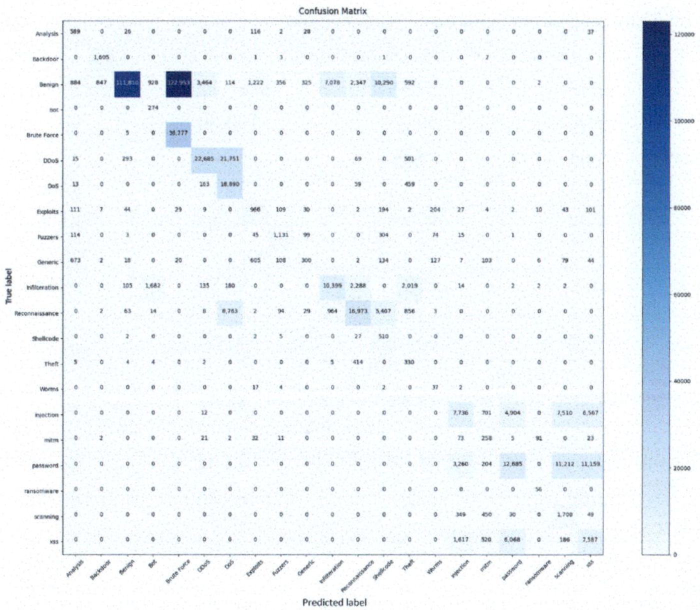

Fig. 4 CM with NF-UQ-NIDS dataset

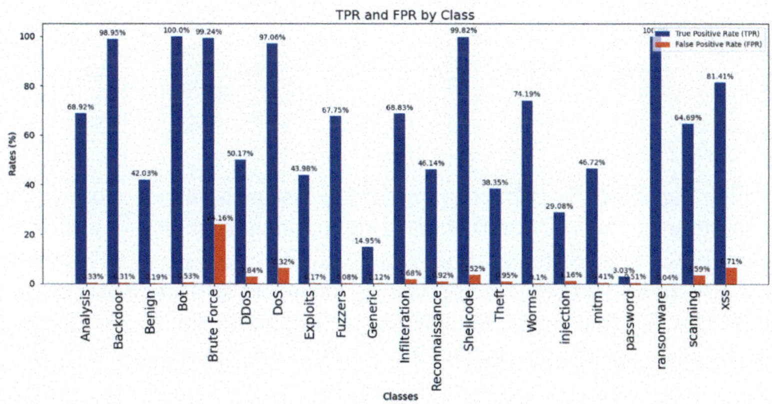

Fig. 5 TPR versus FPR by class

robust ability to accurately identify instances of bot activity. Similarly, "DDOS attack-LOIC-UDP" exhibits high precision (100%), recall (99.80%), and F1-score (99.90%), reflecting the model's effectiveness in detecting UDP-based DDoS attacks. The accuracy analysis shows that the model achieves an average accuracy of 95.69%, demonstrating robust classification capabilities.

Among the classes evaluated, several stand out for their performance values (see Fig. 7). Notably, "Bot" demonstrates an exceptional TPR of 99.98%, indicating the

Table 4 Classification report of NF-CSE-CIC-IDS2018

Label	Accuracy (%)	Precision (%)	Recall (%)	F1-score (%)
Benign	89.33	57.55	46.78	51.61
Bot	99.98	99.33	100.00	99.67
Brute Force-Web	99.64	58.98	58.42	58.70
Brute Force-XSS	99.65	38.26	34.13	36.08
DDOS attack-HOIC	99.97	57.50	66.67	61.74
DDOS attack-LOIC-UDP	100	100	99.80	99.90
DDoS attacks-LOIC-HTTP	93.20	99.98	77.03	87.02
DoS attacks-GoldenEye	88.80	13.53	51.51	21.43
DoS attacks-Hulk	89.46	41.57	47.94	44.53
DoS attacks-SlowHTTPTest	83.69	32.28	85.76	46.90
DoS attacks-Slowloris	97.12	20.00	0.00	0.00
FTP-BruteForce	84.01	66.00	14.23	23.41
Infilteration	99.84	97.60	98.98	98.29
SQL injection	99.98	21.00	100.00	33.33
SSH-BruteForce	89.89	48.84	57.53	52.83
Average	95.69	60.96	55.89	54.45

model's high capability in correctly identifying instances of bot activity, coupled with an impressively low FPR of 0.0095%, highlighting its ability to minimize false alarms. Similarly, "DDOS attack-LOIC-UDP" exhibits a near-perfect TPR of 99.60% with a very low FPR of 0.0055%, underscoring the model's effectiveness in detecting UDP-based DDoS attacks while maintaining a low false positive rate. Furthermore, "Infiltration" showcases a remarkable TPR of 100% and an impressive NPV of 100%.

5.4 CIC-DDoS2019 Dataset

Intrusion detection based on a Multi-layer GraphSAGE model using the CIC-DDoS2019 dataset, comprising 13 classes, reveals varying performance across different classes as illustrated in Table 5 and in the Confusion Matrix in Fig. 8. Notably, "DrDoS-SNMP" demonstrates exceptional precision (95.28%), recall (97.67%), and F1-score (96.46%), indicating the model's robust ability to accurately identify instances of SNMP-based DDoS attacks. Similarly, "TFTP" exhibits relatively high

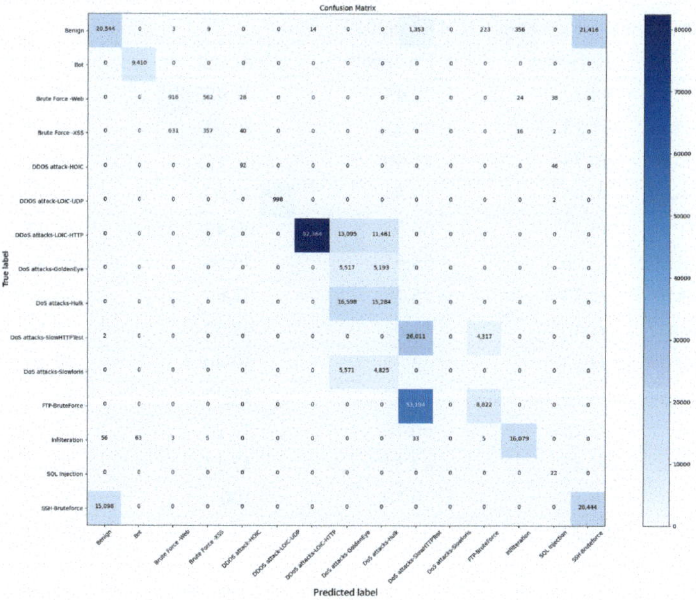

Fig. 6 CM for NF-CSE-CIC-IDS2018

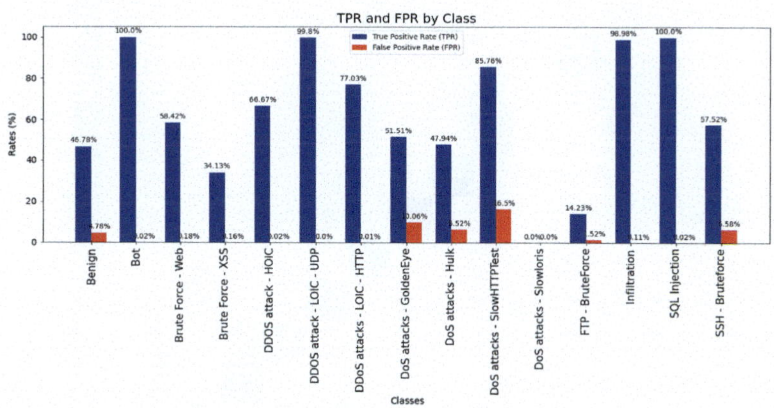

Fig. 7 TPR versus FPR by class

precision (60.76%) and recall (79.22%). With an average accuracy of 94.33%, the model demonstrates efficient classification capabilities.

The class "DrDoS-SNMP" stands out with a high true positive rate (97.07%) and a relatively low false positive rate (5.41%) as shown in Fig. 9, indicating the model's effectiveness in accurately identifying instances of SNMP-based DDoS attacks while minimizing false alarms. This class also demonstrates a high positive predictive value (94.79%), further confirming the model's capability in correctly identifying instances of SNMP-based DDoS attacks among the total predicted positives.

Table 5 Classification Report of CIC-DDoS2019

Label	Accuracy	Precision	Recall	F1-score
BENIGN	98.22	72.43	77.49	74.87
DrDoS_DNS	91.23	49.33	26.64	34.60
DrDoS_LDAP	90.73	41.90	62.18	50.07
DrDoS_MSSQL	89.95	19.84	40.07	60.76
DrDoS_NTP	95.43	64.55	60.41	62.41
DrDoS_NetBIOS	92.96	58.51	76.56	66.33
DrDoS_SNMP	99.35	95.28	97.67	96.46
DrDoS_SSDP	91.30	51.07	68.57	58.54
DrDoS_UDP	95.63	43.85	50.40	46.90
Syn	93.79	76.82	45.47	57.10
TFTP	93.46	60.76	79.22	68.77
UDP-lag	91.87	43.32	37.13	40.00
WebDDoS	98.37	9.50	100	51.00
Average	94.33	52.79	63.26	60.07

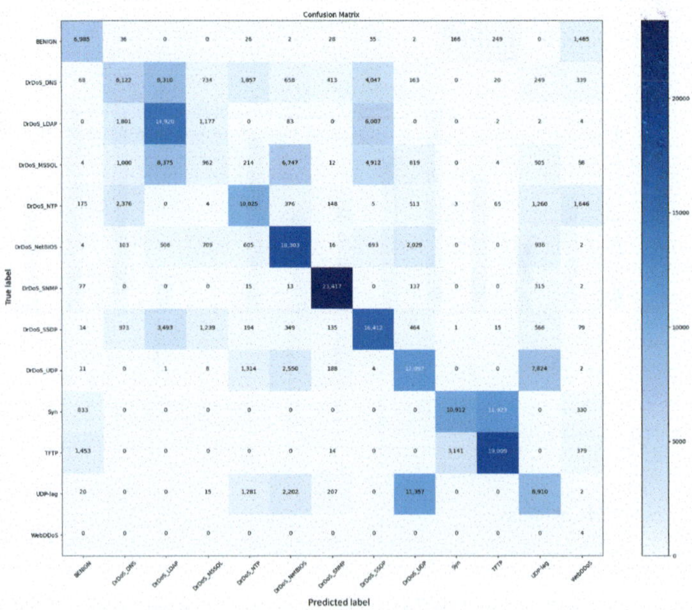

Fig. 8 CM with CIC-DDoS2019

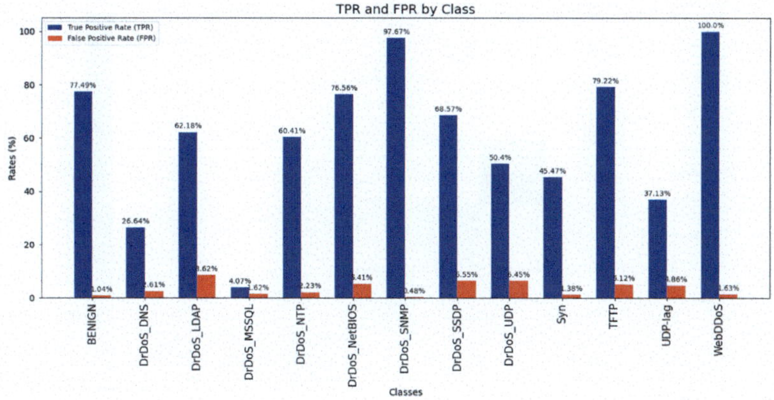

Fig. 9 TPR versus FPR by class

5.5 *InSDN Dataset*

Intrusion detection using a Multi-layer GraphSAGE model on the InSDN dataset reveals varying performance as illustrated in Table 6 and in Fig. 10. The class "DDoS Attack" achieved exceptional precision (99.87%), recall (99.60%), and F1-score (99.73%), indicating the model's high accuracy in identifying instances of DDoS attacks while effectively capturing most actual DDoS attack instances as shown in Table 6. Additionally, the class "DoS" exhibits robust performance with high precision (97.10%), recall (90.40%), and F1-score (93.63%), demonstrating the model's effectiveness in detecting Denial-of-Service attacks. The accuracy analysis shows that the model achieves an average accuracy of 94.68%, demonstrating robust classification capabilities.

Table 6 Classification report of InSDN

Label	Accuracy (%)	Precision (%)	Recall (%)	F1-score (%)
BFA	98.53	18.01	20.26	23.01
BOTNET	95.30	73.00	72.45	44.23
DDoS Attack	99.81	99.87	99.60	99.73
DoS	98.08	97.10	90.40	93.63
Normal	96.32	97.38	83.77	90.06
Probe	86.90	91.32	59.79	72.36
U2R	92.00	50.00	80.00	10.15
Web-Attack	90.53	9.97	87.93	17.42
Average	94.68	67.08	74.28	56.32

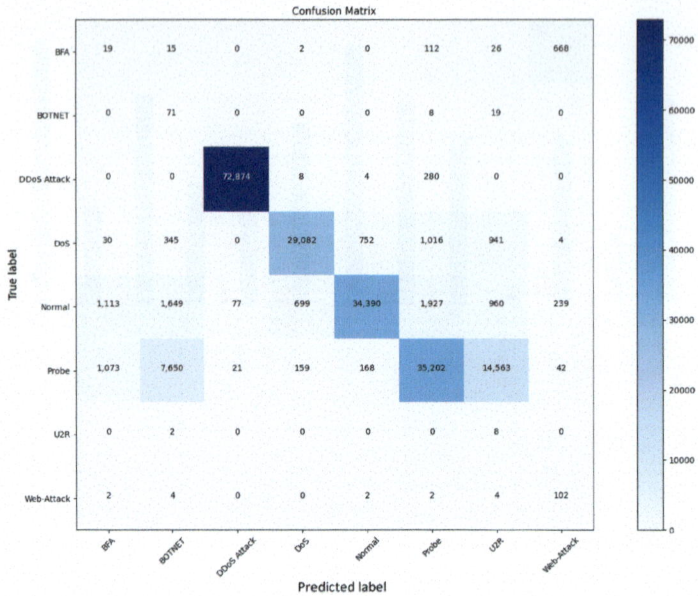

Fig. 10 CM with InSDN dataset

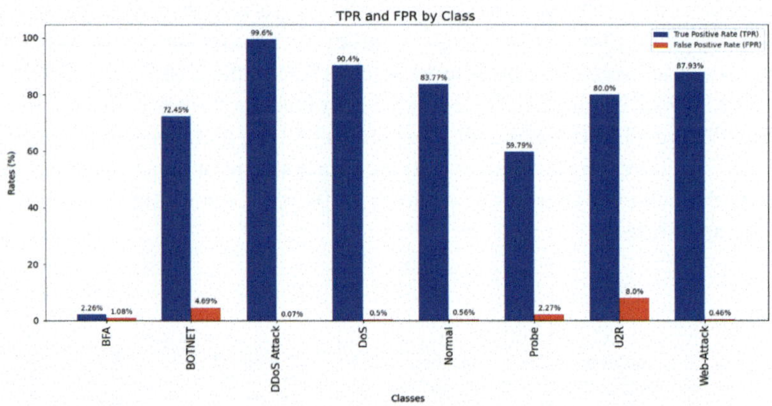

Fig. 11 TPR versus FPR by class

The class "DDoS Attack" is given the best-performing class with respect to both true positive rate (TPR) and false positive rate (FPR) (see Fig. 11). It achieves an impressive true positive rate of 99.61%. Simultaneously, it maintains an exceptionally low false positive rate of 0.03%, signifying the model's capability to minimize misclassifications of non-DDoS attack instances as DDoS attacks. Furthermore, the accuracy for the "DDoS Attack" class is exceptionally high at 99.84%, reinforcing the model's effectiveness in accurately detecting DDoS attacks while maintaining a high level of overall accuracy. Another noteworthy class is “Normal,” which exhibits a relatively high true positive rate of 81.25% and a very low false positive rate of 0.41%.

5.6 *Edge-IIoTset Dataset*

The detailed results of the performance are visually represented in Table 7 below and in the Confusion Matrix in Fig. 12. The performance results of the intrusion detection Edge-IIoT Database Multi-layer GraphSAGE model indicate a strong overall accuracy of 85.96%. The model demonstrates notable precision and recall values for various attack types, with particularly high performance in detecting DDoS-UDP (precision: 99.13%, recall: 99.76%, F1-score: 99.45%) and Normal (precision: 99.58%, recall: 94.38%, F1-score: 96.91%) classes.

Upon analysis of the True Positive Rate (TPR) and False Positive Rate (FPR) (Fig. 13) for Edge-IIoTset, it is evident that the model achieves relatively high TPR values across various attack classes, indicating its capability to effectively detect true positive instances of attacks. For instance, the TPR values for classes such as DDoS-ICMP, DDoS-UDP, MITM, and Vulnerability-scanner, approach or exceed 90%, highlighting the model’s robustness in identifying these specific attack types. Conversely, examining the FPR values reveals the model’s propensity to incorrectly

Table 7 Classification report of Edge-IIoTset

Label	Accuracy (%)	Precision (%)	Recall (%)	F1-score (%)
Backdoor	98.64	43.25	68.19	52.93
DDoS_HTTP	97.71	48.40	27.34	34.94
DDoS_ICMP	99.42	95.35	90.47	94.25
DDoS_TCP	98.95	90.31	59.88	72.01
DDoS_UDP	99.94	99.13	99.76	99.45
Fingerprinting	99.30	20.69	41.33	5.05
MITM	99.98	71.94	100.00	83.68
Normal	95.62	99.58	94.38	96.91
Password	96.54	28.67	35.72	31.81
Port_Scanning	98.41	36.06	73.08	48.29
Ransomware	98.16	7.65	24.82	11.70
SQL_injection	96.87	24.31	16.89	19.93
Uploading	97.58	25.52	22.22	23.76
Vulnerability_scanner	98.02	54.36	76.65	63.61
XSS	58.79	2.55	9.32	4.00
Average	95.59	43.24	50.68	40.00

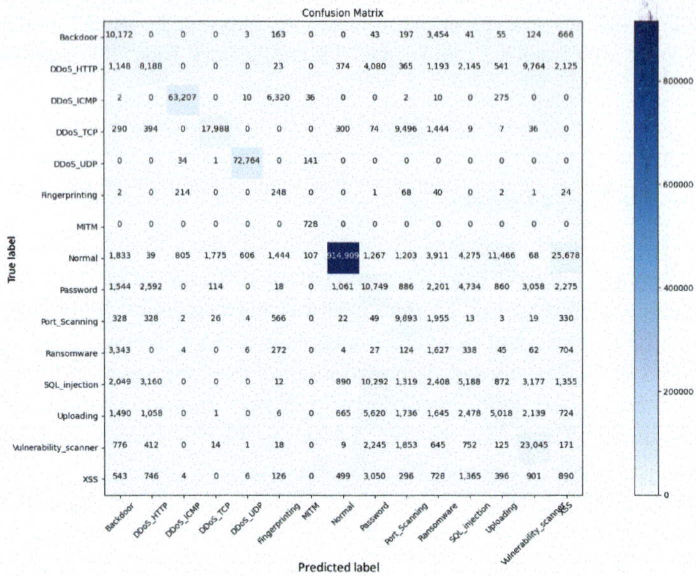

Fig. 12 CM with Edge-IIoTset

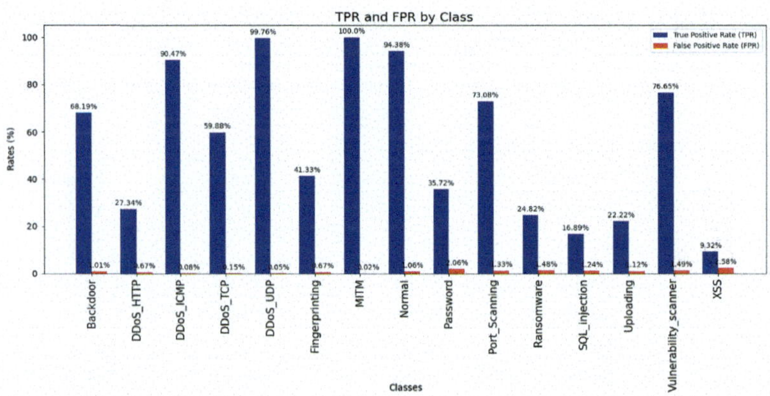

Fig. 13 TPR versus FPR by class

classify normal instances as attacks. While some classes demonstrate relatively low FPR values (e.g., DDoS-TCP, DDoS-ICMP), others exhibit higher rates of false positives (e.g., Fingerprinting, SQL-injection).

5.7 CIC-Darknet2020 Dataset

In the intrusion detection analysis using the Multi-Layer GraphSAGE model on the CIC-darknet2020 dataset, several performance metrics were evaluated for different attack labels as shown in Table 8 and Fig. 14. Among these labels, the "Tor" label stands out with the highest accuracy of 98.23%, indicating that the model accurately classified a significant portion of Tor traffic instances. However, the Precision for Tor traffic is relatively low at 30.85%, suggesting a notable false positive rate. Despite this, the Tor label achieved the highest Recall of 81.77%, implying that the model effectively captured a large proportion of actual Tor traffic instances.

In this scenario (Fig. 15), the TPR values range from 61.15 to 91.75%, indicating the model's effectiveness in capturing true positive instances across different attack categories. The highest TPR of 91.75% suggests that the model successfully

Table 8 Classification report CIC-darknet2020 dataset

Label	Accuracy (%)	Precision (%)	Recall (%)	F1-score (%)
Non-Tor	89.52	93.10	91.75	92.42
Non-VPN	89.39	65.87	61.15	63.42
Tor	98.23	30.85	81.77	44.79
VPN	91.19	68.70	71.76	70.20
Average	92.08	64.63	76.60	67.70

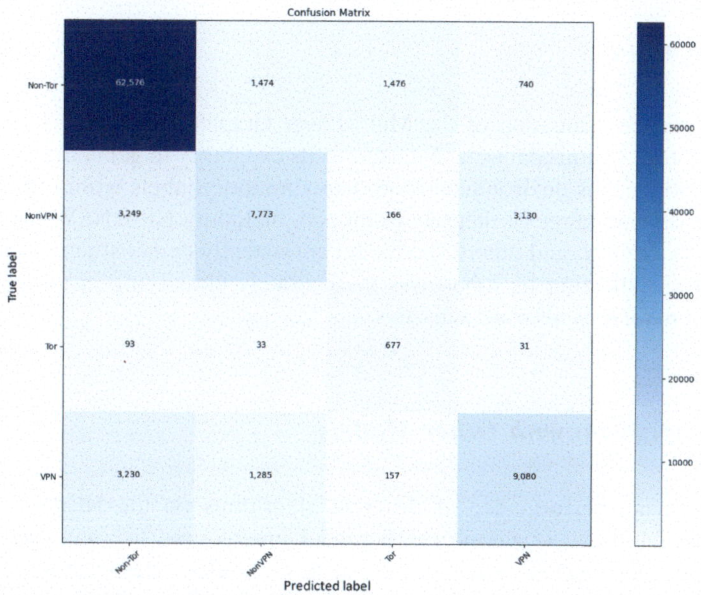

Fig. 14 CM with CIC-darknet2020

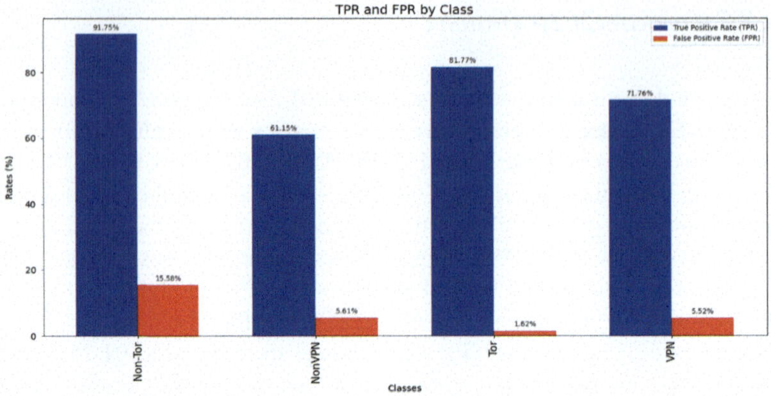

Fig. 15 TPR versus FPR by class

identified a significant portion of positive instances for the corresponding attack category. Conversely, the False Positive Rate (FPR) quantifies the proportion of negative instances that were incorrectly classified as positive by the model. Lower FPR values, ranging from 1.62% to 15.58%, indicate fewer false alarms or instances where benign traffic was mistakenly flagged as malicious. Notably, the lowest FPR of 1.62% corresponds to the attack category with the highest TPR of 91.75%, suggesting a balanced performance in both sensitivity and specificity.

5.8 Discussion

The performance evaluation of the Multi-Layer GraphSAGE model for intrusion detection across various datasets underscores its capability to generalize effectively across heterogeneous environments and detect minority labels within the datasets. Across the diverse range of datasets examined, including NF-UNSW-NB15, CIC-darknet2020, InSDN, and others, the model consistently demonstrates strong accuracy metrics, indicating its robustness in accurately classifying instances of both normal and malicious network activities.

5.9 Comparison with Other Models

In analyzing the performance of different algorithms on the NF-UNSW-NB15 dataset, the Multi-Layer GraphSAGE model emerges as the top performer in terms of both average accuracy and recall in multi-class classification as shown in Table 9. With an average accuracy of 92.34%, the GraphSAGE model significantly outperforms other algorithms such as CNN+LSTM, GAT, and DNN, which

Table 9 Performance results for each algorithm over multiple classifications

Number of clients	Algorithms	Avg. accuracy (%)	Avg. recall (%)
NF-UNSW-NB15 data	CNN+LSTM	67.03	45.55
	GAT	68.44	60.14
	DNN	84.15	60.59
	ML-GraphSAGE (Proposed)	92.34	61.74
CIC-DDoS2019 data	CNN+LSTM	90.38	62.03
	GAT	71.01	56.19
	DNN	91.12	59.03
	ML-GraphSAGE (Proposed)	94.33	63.26

achieve 67.03%, 68.44%, and 84.15% accuracy, respectively. Similarly, on the CIC-DDoS2019 dataset, the Multi-Layer GraphSAGE model continues to showcase its superior performance, achieving an average accuracy of 94.33% and an average recall of 63.26%. This outperforms all other algorithms, including CNN+LSTM, GAT, and DNN, which achieve accuracies of 90.38%, 71.01%, and 91.12%, respectively, and recalls of 62.03%, 56.19%, and 59.03%, respectively.

5.10 Comparison with Previous Studies

In comparing the performance of the Multi-Layer GraphSAGE model with related works and previous studies, particularly in terms of average accuracy and loss across different datasets and models, several insights can be gleaned. Firstly, as illustrated in Table 10, when compared to baseline models such as LSTM and SVM-IDS on the CIC-DDoS2019 dataset, the Multi-Layer GraphSAGE model achieves higher average accuracy, outperforming LSTM Baseline (94.08%) and SVM-IDS (94.02%) with an accuracy of 94.33%. Furthermore, compared to the Enhanced GraphSAGE model on the CICIDS2017 dataset, the Multi-Layer GraphSAGE model maintains competitive performance in terms of average accuracy. While the Enhanced GraphSAGE achieves an accuracy of 93.32%, the ML-GraphSAGE model achieves a slightly higher accuracy of 94.33% on the CIC-DDoS2019 dataset.

6 Conclusion

This paper presents a comprehensive investigation into the application of a Multi-Layer GraphSAGE model for intrusion detection across diverse network environments. Through extensive experimentation on multiple datasets, the proposed

Table 10 The comparison of the proposed model to the state-of-the-art results

Authors/year	Datasets	Model	Classification	Accuracy (%)	Loss
Devrim Akgun et al. (2022) [24]	CIC-DDoS2019	LSTM baseline	Multi (13)	94.08	5.92
ZIHAN WU et al. (2022) [25]	CIC-DDoS2019	SVM-IDS	Binary	94.02	5.98
Zhong-Dang Le et al. (2023) [26]	CICIDS2017	Enhanced GraphSAGE	Multi (3)	93.32	6.68
Jamal Hussain et al. (2012) [27]	UNSW-NB15	DNN-SAE	Multi (10)	81.70	18.30
Alamri and Thayananthan (2020) [28]	InSDN	Extreme gradient boosting algorithm	Multi (8)	91.26	–
This paper	CIC-DDoS2019	ML-GraphSAGE	Multi (13)	94.33	5.67
	InSDN		Multi (8)	94.68	5.32
	Edge-IIoTset		Multi (15)	95.59	4.41

GraphSAGE model consistently demonstrates superior performance in accurately identifying various types of network intrusions. Multi-Layer GraphSAGE shows superior average accuracy across diverse datasets, often outperforming baselines and related models. Future work can optimize hyperparameters to enhance intrusion detection capabilities in varied network environments.

References

1. Thiyam B, Dey S (2023) Efficient feature evaluation approach for a class-imbalanced dataset using machine learning. Procedia Comput Sci 218:2520–2532
2. Hussain J, Hnamte V (Sep 2021) A novel deep learning-based intrusion detection system: software defined network. In: 2021 international conference on innovation and intelligence for informatics, computing, and technologies (3ICT). IEEE, pp 506–511
3. Pujol-Perich D, Suárez-Varela J, Cabellos-Aparicio A, Barlet-Ros P (2022) Unveiling the potential of graph neural networks for robust intrusion detection. ACM SIGMETRICS Perform Eval Rev 49(4):111–117
4. Layeghy S, Portmann M (2022) On generalisability of machine learning-based network intrusion detection systems. arXiv:2205.04112
5. Khraisat A, Gondal I, Vamplew P, Kamruzzaman J, Alazab A (2019) A novel ensemble of hybrid intrusion detection system for detecting internet of things attacks. Electronics 8(11):1210
6. Li X, Xu M, Vijayakumar P, Kumar N, Liu X (2020) Detection of low-frequency and multi-stage attacks in industrial internet of things. IEEE Trans Veh Technol 69(8):8820–8831

7. Süzen AA (2021) Developing a multi-level intrusion detection system using hybrid-DBN. J Ambient Intell Humlzed Comput 12(2):1913–1923
8. Liang W, Li KC, Long J, Kui X, Zomaya AY (2019) An industrial network intrusion detection algorithm based on multifeature data clustering optimization model. IEEE Trans Ind Inform 16(3):2063–2071
9. Lan J, Lu JZ, Wan GG, Wang YY, Huang CY, Zhang SB, Ma JN et al (2022) E-minbatch graphsage: an industrial internet attack detection model. Secur Commun Netw 2022
10. Friha O, Ferrag MA, Shu L, Maglaras L, Choo K-KR, Nafaa M (2022) FELIDS: federated learning-based intrusion detection system for agricultural Internet of Things. J Parallel Distrib Comput 165:17–31
11. Lo WW, Layeghy S, Sarhan M, Gallagher M, Portmann M (Apr 2022) E-GraphSAGE: a graph neural network based intrusion detection system for IoT. In: NOMS 2022-2022 IEEE/IFIP network operations and management symposium. IEEE, pp 1–9
12. Cheng Q, Wu C, Zhou S (2021) Discovering attack scenarios via intrusion alert correlation using graph convolutional networks. IEEE Commun Lett 25:1564–1567
13. Caville E, Lo WW, Layeghy S, Portmann M (2022) Anomal-E: a self-supervised network intrusion detection system based on graph neural networks. Knowl-Based Syst 258:110030
14. Zhou J, Xu Z, Rush AM, Yu M (2020). Automating botnet detection with graph neural networks. arXiv:2003.06344
15. Li J, Pang G, Chen L, Namazi-Rad MR (Oct 2023) HRGCN: heterogeneous graph-level anomaly detection with hierarchical relation-augmented graph neural networks. In: 2023 IEEE 10th international conference on data science and advanced analytics (DSAA). IEEE, pp 1–10
16. Chang L, Branco P (2021) Graph-based solutions with residuals for intrusion detection: the modified E-GraphSAGE and E-ResGAT algorithms. arXiv:2111.13597
17. Hamilton W, Ying Z, Leskovec J (2017) Inductive representation learning on large graphs. Advances in neural information processing systems, 30
18. Sarhan M (2020) Netflow datasets. http://staff.itee.uq.edu.au/marius/NIDSdatasets
19. Sharafaldin I, Lashkari AH, Ghorbani AA (2018) Toward generating a new intrusion detection dataset and intrusion traffic characterization. ICISSp 1:108–116
20. Sharafaldin I, Lashkari AH, Hakak S, Ghorbani AA (Oct 2019) Developing realistic distributed denial of service (DDoS) attack dataset and taxonomy. In: 2019 international Carnahan conference on security technology (ICCST). IEEE, pp 1–8
21. Elsayed MS, Le-Khac NA, Jurcut AD (2020) InSDN: a novel SDN intrusion dataset. IEEE Access 8:165263–165284
22. Habibi Lashkari A, Kaur G, Rahali A (Nov 2020) Didarknet: a contemporary approach to detect and characterize the darknet traffic using deep image learning. In: Proceedings of the 2020 10th international conference on communication and network security, pp 1–13
23. Ferrag MA, Friha O, Hamouda D, Maglaras L, Janicke H (2022) Edge-IIoTset: a new comprehensive realistic cyber security dataset of IoT and IIoT applications for centralized and federated learning. IEEE Access 10:40281–40306
24. Akgun D, Hizal S, Cavusoglu U (2022) A new DDoS attacks intrusion detection model based on deep learning for cybersecurity. Comput Secur 118:102748
25. Wu Z, Zhang H, Wang P, Sun Z (2022) RTIDS: a robust transformer-based approach for intrusion detection system. IEEE Access 10:64375–64387
26. Zhong M, Lin M, He Z (2023) Dynamic multi-scale topological representation for enhancing network intrusion detection. Comput Secur 135:103516
27. Hussain J, Hnamte V (Aug 2021) Deep learning based intrusion detection system: software defined network. In: 2021 Asian conference on innovation in technology (ASIANCON). IEEE, pp 1–6
28. Alamri HA, Thayananthan V (2020) Bandwidth control mechanism and extreme gradient boosting algorithm for protecting software-defined networks against DDoS attacks. IEEE Access 8:194269–194288

SlowROS: Investigating the Possibility to Perpetrate Slow DoS Attacks Against Robot Operating System

Enrico Cambiaso, Luca Durante, and Ivan Cibrario Bertolotti

Abstract Properly protecting industrial environments is a relevant activity in the network security field, due to the significant impact that potential cyberattacks may have. By exploiting the possibility of designing new threats targeting industrial networks, in this paper we present SlowROS, a novel Slow DoS Attack targeting the Robot Operating System, ROS. In our work, we validated the threat through the execution of dedicated test sets on a real network, composed of multiple interacting components. Obtained results show that the proposed attack leads to a denial of service against ROS components, making them unable to operate correctly. Indeed, although we did not directly targeted it, we also observed the attack impact on a typical ROS device and discussed how a production pipeline can be blocked by SlowROS. In addition, we found non-negligible peaks of data exchanged, thus we discussed how it is possible to optimize the threat and minimize its impact on the network in terms of network bandwidth needed by the attacker, by exploiting the potentially endless connection timeout of the targeted service.

Keywords Cybersecurity · Denial of service · Slow dos attack · Industrial security · ROS · Critical infrastructure protection

E. Cambiaso · L. Durante · I. C. Bertolotti (✉)
Consiglio Nazionale Delle Ricerche (CNR), IEIIT Institute, Torino, Italy
e-mail: ivancibrario.bertolotti@cnr.it

E. Cambiaso
e-mail: enrico.cambiaso@cnr.it

L. Durante
e-mail: luca.durante@cnr.it

© The Author(s), under exclusive license to Springer Nature Switzerland AG 2025

S. Femmam and P. Lorenz (eds.), *Recent Advances in Communication Networks and Embedded Systems*, Lecture Notes on Data Engineering and Communications Technologies 256, https://doi.org/10.1007/978-3-031-92449-1_9

1 Introduction

Protecting industrial and Operational Technology environments is a primary goal for critical infrastructure protection [8]. In this context, Robot Operating System (ROS) [25] is a set of open-source libraries and tools that provide commonly used robot control functionalities and algorithms, along with a high-level software packaging framework that encourages modularity and sound software engineering practices. Moreover, ROS provides hardware abstraction and platform-independent robotic device control and access. Data exchange among system components is implemented by means of a pervasive message passing subsystem based on the publisher–subscriber communication paradigm.

Even though its initial target audiences were mainly robotic research and hobbyists' projects [40], ROS gained more and more industrial interest with time, thanks to initiatives like the ROS Industrial Consortium [29], whose members include leading robot OEM manufacturers, end-users such as large international manufacturing companies, and research institutes worldwide. Our research work has been focused on the most recently released version of ROS 1 (Noetic Ninjemys). This version is meant to be the last version of ROS 1 before the transition to ROS 2, slated for mid-2025. Although ROS 2 introduces novel concepts to the protocol, nowadays, it still makes sense to assess and analyze the security of ROS 1, because its installed base in real-world applications is today quite significant. By contrast, ROS 2 is still in active development and some of its features are not yet finalized. Moreover, the final phase-out of ROS 1 systems and their replacement with ROS 2 will undoubtedly extend significantly after 2025, considering the typical inertia of industrial equipment refurbishment, thus making this activity still significant for years to come.

Regarding security aspects of ROS-powered networks and, more in general, industrial contexts, threats are not new. While Stuxnet represents the most known attack, injecting malicious components on nuclear plants [32], ROS has also been the target of different threats, such as data manipulation [5], ransomware [18], and denial of service (DoS) [9]. Considering the last-mentioned menace,

Slow DoS Attacks [7] are emerging, last-generation denial of service threats. They aim at covertly exhausting some kind of resource of a victim component or subsystem within the targeted system, thus compromising its ability to service legitimate requests from other components and affecting the overall availability of the system to its intended users. As, to the best of our knowledge, Slow DoS Attacks targeting ROS are not found in literature, by following the approach to port such threats to the industrial context, in this paper we aim at evaluating the possibility to perpetrate Slow DoS Attacks against a specific, mandatory ROS component—the ROS Master. Given the limited attack requirements characterizing slow DoS threats, in conjunction with the difficulties of detecting ongoing threats [23], we believe it is important to investigate the possibility of injecting such attacks in a ROS-powered industrial network. While studying this, we designed SlowROS, an innovative slow DoS cyber-attack specifically designed to target ROS services. To the best of our knowledge, SlowROS represents the first Slow DoS Attack designed to target industrial environments. Moreover, by evaluating its performance, we found that the targeted system makes use of extremely high connection timeouts: As such timeouts can be misused by a slow DoS attacker, they should be considered as an exploitable vulnerability.

The remainder of the paper is organized as follows. After providing an overview of the state of the art and related work in Sect. 2, the paper presents the proposed Slow DoS Attack against ROS in Sect. 3. Then, Sect. 4 describes the test bed used to demonstrate it, while Sect. 5 presents and discusses experimental results. Finally, Sect. 6 concludes the paper and outlines further work on the topic.

2 Related Work

From its inception, and especially since its application domain spread from purely academic research into the industrial domain, ROS security has been the subject of a fair amount of scrutiny by researchers and practitioners alike.

The authors of [15] were among the first to study ROS 1 vulnerabilities and related countermeasures. In fact, they highlighted that there is neither enforceable authentication nor encryption between ROS nodes and the ROS Master itself is susceptible to replay attacks and service hijacking. Similar considerations, along with a more in-depth analysis, can be found in [1, 31] as well. In their work, these authors did not focus on DoS attacks specifically and did not provide any details about their execution to target ROS. Later on, other researchers conducted penetration testing on ROS systems [11] and were able to demonstrate more elaborate attacks [4]. Namely, they were able to crash or disable some ROS nodes, by running an exploit-based attack to make them unavailable to other nodes, and hence, make the ROS system as a whole inoperable. While the attacks lead to similar results compared to the threat proposed in this paper, in their case the attack strategy does not involve the use of "slow" approaches, in favor of methodologies exploiting specific software vulnerabilities (e.g., CVE exploits). Similarly, a recent survey [21] provides a comprehensive review of ROS security, but still does not mention Slow DoS Attacks. Last, the official "Python library reference" [30] warns that the `xmlrpc.server` module—used by the ROS Master as XML-RPC layer—is not robust against *maliciously crafted* data. However, it does not mention the possibility of perpetrating DoS attacks by means of *legitimate* data, as we do in the present work. Furthermore, although there are currently about 100 known security vulnerabilities and CVEs against `ros-comm` (the ROS communication component) [20], to the best of the authors' knowledge none of them discusses DoS attacks of the kind presented in this paper.

For what concerns detection and mitigation strategies, [22] proposed to improve ROS system security by integrating ROS with Message Queuing Telemetry Transport (MQTT) [41], a widely used Internet of Things (IoT) protocol. They consider DoS only in passing, without providing any specific countermeasures to prevent it. In addition, the MQTT protocol itself has recently been found to be vulnerable to some kinds of DoS attack [37]. More recent work focused on detecting attacks against ROS by means of machine learning algorithms [3, 42]. However, [3] considers only data injection attacks, while [42] takes into consideration only traditional DoS attacks, in which large amounts of illegitimate data are involved. Moreover, [28] advocates the use of deep reinforcement learning to identify vulnerabilities in ROS-based robot navigation, but without considering DoS attacks explicitly. The ROS documentation [25] also provides several recommendations on how to improve security. The first one is to restrict access to the ROS network by isolating it or by interposing a firewall. Although valuable, this recommendation is hard to enforce and guarantee, especially on production environments. In fact, a study [10] found a plethora of ROS systems that were exposed to the public Internet and it is unclear whether the situation has improved since then or not. Moreover, the ROS documentation advocates the adoption of additional security-oriented components, namely, Rosbridge and SROS. Rosbridge is a WebSocket–ROS bridge that grants web applications controlled access to ROS nodes. Provided the WebSocket interface is secure, it is useful to prevent attackers from reaching the ROS system from the Internet. Nevertheless, it may be ineffective against insider threats, which act from the private, industrial network rather than from the Internet. Also, the WebSocket interface may be vulnerable to specific CVEs. By contrast, SROS enhances ROS security by adding TLS support to the existing ROS protocols. However, at the time of this writing, its documentation says the component is highly experimental and under heavy development, which casts doubts about its ability to be deployed in a production environment. The final recommendation provided in the ROS documentation is to switch to ROS 2, which relies on the Data Distribution Service (DDS) standard [24] for communication. Besides the reasons for focusing on ROS 1 for the time being, already mentioned in the introduction, it is worth remarking that significant vulnerabilities [34] have recently been found for DDS, too, leading to an ICS advisory [2].

By considering the threats mentioned, to the best of our knowledge, the effective implementation of a Slow DoS Attack targeting ROS is not found in literature. Regarding the discussed protection approaches, while some methodologies may limit network access, they are not designed to protect from specific DoS attacks. In this context, existing works focus on categorization of such threats [7], optimization [35], exploitation of specific environments such as Internet of Things [16, 36] or SDN [27]. In addition, several works attempt to detect Slow DoS Attacks by using different identification techniques, such as statistics [12], machine learning [14], or deep learning [13]. Other works overview detection approaches [38]. To the best of our knowledge, the arena of Slow DoS Attacks does not include any threats specifically designed against ROS. Therefore, targeting a ROS-equipped environment through a Slow DoS Attack enables the possibility to extend the scope of such threats to also include industrial environments. Considering the multitude of threats designed to target conventional Internet services [19], the investigation of a potential application of specific categories of threats in a different context opens to new potential research challenges to address.

3 SlowROS Attack Description and Analysis

ROS systems work, for the most part, according to the publisher–subscriber paradigm. In particular, both publishers and subscribers use HTTP REST-based APIs including XML-formatted data for all their interactions with the ROS Master, which acts as a centralized agent, or broker, for topic registration and lookup. The XML-RPC [39] protocol they use is a connection-oriented protocol that issues HTTP POST requests to transport XML-encoded RPC requests and receive responses. As usual, HTTP traffic is delivered by TCP. Importantly, publishers and subscribers initiate TCP connections (acting as clients), which are then accepted by the ROS Master (also denoted as *server* in the following). This makes these interactions similar to those with a web server or a MQTT broker and, in turn, could make the ROS Master susceptible to a set of cyber-attacks affecting network services in the IT domain.

Considering such aspects, the Slow DoS Attack proposed in this paper, SlowROS, is going to act in two stages: (i) at first, a specific amount of simultaneous connections with the server is established, in order to seize all available connection slots the (application) server is able to support. Then, (ii) each connection is kept alive as long as possible, in order to prevent its closure by the server.

Such typical behavior of Slow DoS Attacks [7] may lead to a denial of service on the server, also attempting to sustain such a state for as long as possible. In detail, the number of connections the attacker has to simultaneously establish and keep alive depends on the application server configuration. Although for specific servers, such as the Apache web server, such value can be set up, it may not have an impact on the attack success, while, by increasing such value, the attacker would require more resources to accomplish its goal. Instead, in order to keep the connection open as long as possible, different approaches are possible: In our case, by implementing a so-called *Long Request DoS* [7], we divide the transmission of the request payload into multiple cycles, by sending a predefined payload to the server and by splitting it to send just a portion of it at each round. In particular, the considered payload sent by the attacker may vary and may exploit different functionalities/methods provided by the ROS server. In our case, we adopted a payload similar to the one reported in Fig. 1.

For the sake of readability, in the figure, we have indented content, while, when sending data over the network, content is not indented. Also, we tagged specific information (such as the value of the HTTP Host parameter), to provide a more general payload applicable independently of the target. Finally, although not shown in the figure to improve readability, it is worth mentioning that newlines in the header section are represented by `\r\n` characters, while, in the XML-RPC section of the payload, they just include `\n` characters. Such implementation depends on how the different sections of the payload are interpreted by the server.

It should also be noted that, although we adopted the specific payload reported in Fig. 1, any payload compliant to the protocol is valid to perpetrate a successful SlowROS attack. In addition, payloads not compliant with the ROS protocol may

Fig. 1 Payload content sent by SlowROS to the ROS publisher. HTTP headers are shown in red, XML-RPC encapsulation elements in green, and ROS-specific fields in blue

```
POST /RPC2 HTTP/1.1
Host: {ros_server_host}:11311
Accept-Encoding: gzip
Content-Type: text/xml
User-Agent: Python-xmlrpc/3.8
Content-Length: {payload_length}

<?xml version='1.0'?>
<methodCall>
  <methodName>registerPublisher</methodName>
  <params>
    <param>
      <value><string>/talker_{id}</string></value>
    </param>
    <param>
      <value><string>/chatter</string></value>
    </param>
    <param>
      <value><string>std_msgs/String</string></value>
    </param>
    <param>
      <value><string>http://{endpoint}/</string></value>
    </param>
  </params>
</methodCall>
```

lead to a successful DoS as well, although they could trigger an early disconnection by the server (e.g., raising HTTP 500 errors when malformed data is received).

As the payload sent by the attacker is a legitimate payload compliant with the ROS protocol, by analysing the data exchanged by the client and the server, it is not trivial to detect an ongoing attack. In addition, although in this paper we consider, for simplicity, a single attacking host, whose detection of the malevolent aim at the network layer would be trivial, a real attacker may distribute the threat to multiple malicious nodes, coordinated to perpetrate a common attack against the same target. In such a case, detection of an ongoing attack would be more challenging.

In order to evaluate the possibility to carry out a successful attack, also as a function of the payload sent and the attack approach adopted, in terms of timings and data sent at each round of attack, in the next section we are going to define the test bed considered in our experimentation.

4 Test Bed

In order to validate the efficacy of the threat introduced in Sect. 3, we built a setup representing a full ROS-powered network. Overall, the test bed being presented and used to conduct and analyze the Slow DoS Attack is meant to represent a simplified, but still realistic, implementation of a typical industrial production line. As shown in Fig. 2, it consists of four nodes interconnected by means of a 100BaseTX shop-floor network, depicted as a green line. This network emulates a local plant network. More specifically:

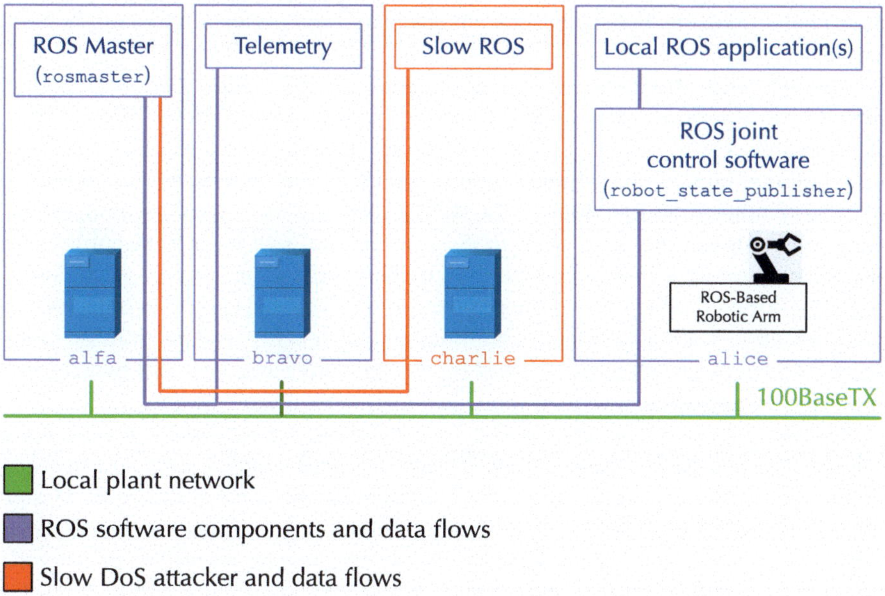

Fig. 2 ROS test bed for Slow DoS Attacks

- Node `alice`, depicted on the right of the figure, is an Interbotix PincherX-100 robot arm (4 DOF plus gripper) controlled by an 8th Gen Intel NUC equipped with the Interbotix-provided distribution of Ubuntu 20.04 and ROS `noetic`. The NUC runs the ROS joint control software and, possibly, additional ROS applications that implement any local robot arm control algorithms needed by the application at hand.
- Nodes `alfa`, `bravo`, and `charlie` are based on Raspberry PI 3 model B boards, also equipped with an off-the-shelf distribution of Ubuntu 20.04 and ROS `noetic`. Nodes `alfa` and `bravo` run ROS code that performs various higher-level system coordination and supervision functions. According to the guidelines presented in [17], `alfa` also hosts the centralized ROS Master (listening on the default port 11311), which all other ROS nodes need to access for topic registration and lookup. Node `charlie` is the malicious node that will conduct the Slow DoS Attack.

In a real industrial system, the network shown in the figure would invariably include a gateway that grants external access to it, for instance, through a wired or Wi-Fi connection to a plant-wide SCADA or MES network. This node would usually incorporate a firewall that is supposed to protect the shop-floor network from unauthorized access, while still allowing control and supervision traffic to pass through. In this paper, we assume a remote attacker somehow reached node `charlie` (or connected it to the network) and injected malicious software into it. This assumption does not hinder the realism of our scenario, as the attacker may, for instance, exploit specific vulnerabilities to accomplish remote command execution or command injection, thus bypassing any gateway-level protection that might be in place.

In the specific test bed version used in this paper, the ROS Master has been implemented by means of the off-the-shelf `rosmaster` process found in the ROS `noetic` distribution. The joint control software running on `alice` is the `robot_state_publisher` application provided by Interbotix with the robot arm, whose communication functions to the ROS Master are also based on unmodified, lower-level software libraries from ROS `noetic`. The high-level ROS application (also running on `alice`) is simulated by a Python application, derived from one of the ROS examples, which drives the gripper through a closed Cartesian trajectory in space indefinitely. Finally, system coordination and supervision functions are simulated by a telemetry application based on the standard `rostopic` tool, which runs on `bravo` and periodically reports joints position, velocity, and torque.

This rather conservative approach is meant to realize a test bed that represents a vulnerable industrial system and guarantees that neither application-specific, nor vendor-specific code is involved in the communication between the ROS Master and the rest of the ROS software. Although our test bed represents a scaled-down version of a real plant, its network protocols and architecture, as well as its network equipment, are still equivalent to a production scenario. Being such components the target of our study, our test bed should therefore be considered realistic. This aspect is crucial to ensure that the attack discussed in Sect. 3 does not depend on the specific test bed in use or any custom software that runs on it, and can hence be exploited on a variety of ROS-based systems.

5 Tests and Results

In this section of the paper, we detail the tests we have performed to validate the SlowROS attack and we discuss obtained results. Particularly, in order to perpetrate a Slow DoS Attack, the first action needed is to assess the timeout used by the server to close pending connections. Hence, by using the identified value to define the *Wait Timeout* of the attack [7], we will evalute the success of SlowROS against our target service.

5.1 Identification of the Server Timeout

As anticipated, to evaluate the success of the proposed threat, first of all the value of the Wait Timeout [7] adopted by the server must be identified. In fact, servers are typically programmed to close pending connections after a specific timeout expiration. The identification of such timeout represents a relevant activity for a Slow DoS attacker, as it is possible to tune the attack to optimize the resources of the adversary, to wake up network connections in proximity of (just before) the expiration of the timeout. Such approach would keep connections alive without experiencing closures, thus denying legitimate users the possibility to successfully connect to the service.

To identify the timeout adopted by the server, which typically assumes a static value and does not depend on the load of the server, establishing just a single connection with the server and keeping it alive without sending any data is sufficient. Sooner or later, the server may interrupt the communication. The time T_{max} elapsed between the last packet sent by the attacker and the time of the first RST/FIN packet sent by the server identifies the first value of Wait Timeout which is not valid for our scenario. Therefore, our attack will adopt a Wait Timeout value $T_{WT} < T_{max}$.

Considering this, by running the tests, we have established a single TCP connection between our attacker, `charlie`, and the ROS Master service, running on `alfa` on port 11311. Our connection was programmed to self-close after $T = 3600$ s, the duration of our observation, if the server does not close it beforehand. We therefore expect to find a value of $T_{max} \in [0, T]$. Particularly, by running our test, we found $T_{max} = 3600.001239$ s. Although such value is greater than the maximum connection duration T, it should be noted that the additional time depends on how the client OS schedules connection-related activities for processes being terminated.

By considering obtained results, we observed that no server-side connection closures are potentially occurring. Such a result is particularly relevant, as we identified a potential weakness malicious users may exploit. Although further tests may focus on extending T, in case of Slow DoS Attacks, where even having a timeout equal to 60 s has a significant impact on the reduction of the attacker's bandwidth, our results assume particular importance, as we found that the server is potentially not closing idle connections for at least 3600 s. Considering that, by defining $T_{max} = 3600$ s to simplify tests execution and stick with test cases lasting at most one hour, although a slight increase of the bandwidth is required by the attacker, compared to a higher value approaching T_{max}, we adopt in the following tests $T_{WT} = 600$ s, which is lower than the T_{max} value we found.

5.2 *SlowROS Attack Execution*

After the identification of the Wait Timeout $T_{WT} = 600$ s, our aim is to evaluate the possibility to successfully perpetrate a SlowROS attack against our target system. In order to do that, as outlined in Sect. 3, we need to set up a specific amount N of concurrent connections to establish with the server and keep alive as long as possible. By considering that such value depends on the targeted server [6], as a first approach, we adopted $N = 2000$. Although such value may seem particularly high, it potentially allows us to define a top baseline to adopt when executing SlowROS. In addition, it should be considered that, in case the number of connections accepted by the server N_s is lower than N, the "extra" connections (from $N_s + 1$ to N) will be in idle state, not managed by the server.

Considering our attack setup, according to Sect. 4, we have observed the behavior of our target system for a period $T = 3600$ s. We have considered a SlowROS attack

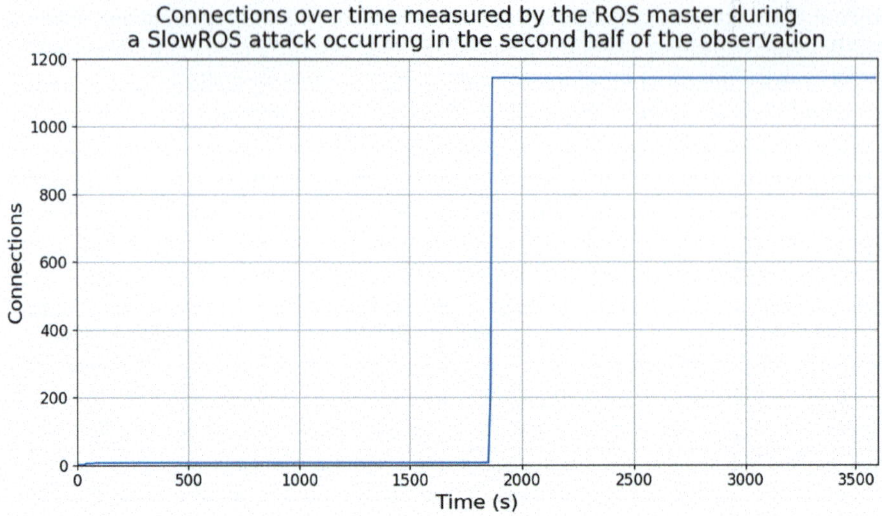

Fig. 3 Effects of a SlowROS attack measured on a running ROS Master node for a $T = 3600$ s observation period. The attack is perpetrated in the $[\frac{T}{2}, T]$ time window

executed in the time window $[\frac{T}{2}, T]$ Hence, while the first half of our observation $([0, \frac{T}{2}) = [0, 1800))$refers to a situation without any attack in place, the second half $([\frac{T}{2}, T] = [1800, 3600])$ includes a SlowROS threat in our target network, for its entire duration. During our test, we analyzed the system behavior, by monitoring, second by second, the number of connections managed by the server. Also, during the test, we captured telemetry information of the robotic arm `alice`, to retrieve and share real-time joint motion information.

Figure 3 shows the effects on the server of a SlowROS attack. It can be noted that the peak on the number of connections occurs in the second half of our observation. We found the maximum number of connections accepted by the server is $N_s < N$ and, in particular, $N_s = 1143$. Therefore, an execution of the attack using $N = N_s = 1143$ would probably have led to similar results.[1] In addition, we measured that all the N_s connections the server is able to manage was established within $\simeq 7$ s since the beginning of the attack. Therefore, our SlowROS threat has been able to lead to a denial of service state on the victim in just 7 s; such state is maintained for the entire attack duration.

During our test, we also measured that the overall amount of data exchanged with the attacker is equal to 622.063 bytes. In detail, while, on average, an attack bandwidth equal to 2.765 bps is required to the attacker, as the attack is characterized by bursts occurring during the so-called slow sending phase [7], we have measured a peak bandwidth of 1.22 Mbps. Although such a value may seem high, considering

[1] Typically, a lower number may be required, to consider, e.g., always present and already established connections used for logging purposes.

that a "slow" DoS attack is performed, it should be considered that it represents the maximum amount, while most of the time, 98.94% of the time in particular, not even a single packet is exchanged with the server. In addition, in view of a potential optimization of the attack and a consequent reduction of the network bandwidth associated to such malicious activity, it should be noted that the T_{WT} value in use is lower than the timeout adopted by the server: using a higher value, in combination with different sending strategies, may lead to reduced bandwidth peaks. Finally, by considering a lower N value, approaching N_s, it should be possible to reduce the attack bandwidth, although we do not expect relevant decreases. The execution of such additional tests may be scope of further work on the topic.

Considering our attack execution, by analysing telemetry data collected from the `bravo` host, referring to `alice` (the robotic arm) movements, we found that the robot joints are moving in normal conditions, by repeating predefined patterns over time. Instead, joint positions no longer change during the attack, hence proving that the robot became motionless.

To better show the effects of the attack on the robotic arm, Fig. 4 shows how telemetry data change over time during our observation period of $T = 3600$. For the sake of readability, the figure only shows data related to three joints, identified as *shoulder*, *elbow*, and *wrist_angle*. As it is possible to notice, for the first half of the capture, the robotic arm is repeatedly moving, to accomplish its tasks. After the begin of the attack, all shown joint positions become static.

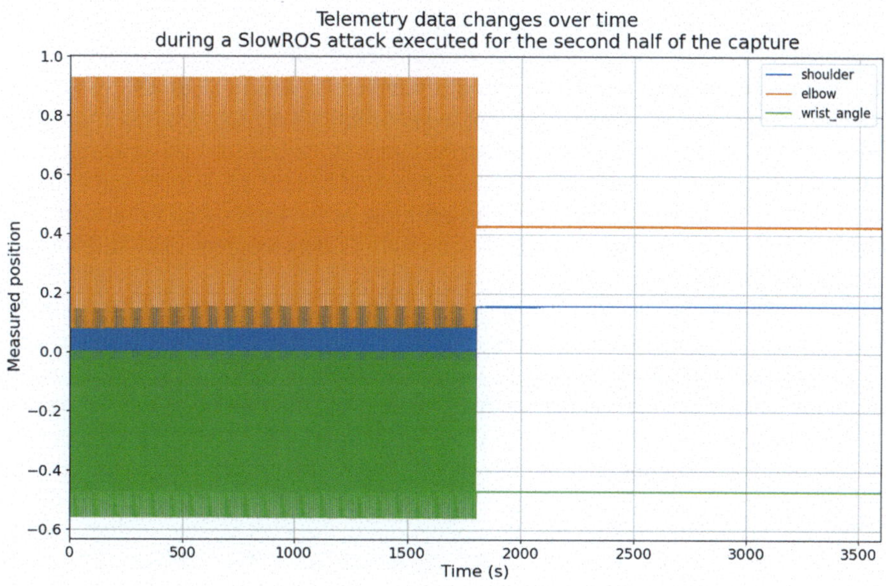

Fig. 4 Telemetry data variations over time observed for $T = 3600$ s. A malicious SlowROS attack is perpetrated in the $[\frac{T}{2}, T]$ time window

Such behavior demonstrates on one hand that SlowROS successfully led to a DoS on the ROS service. On the other hand, the proposed threat shows how it is able to interrupt working machines in the targeted environment. Considering the costs of such interruption on a real production environment, the attack should be considered a serious threat targeting industrial contexts.

6 Conclusions and Future Works

Scrutinizing the security of industrial communication protocols is a crucial activity in the network security field, as the impact of potential malfunctioning may cause financial losses to the targeted organization or even compromise employees and citizen safety. In this context, we have focused on Robot Operating System (ROS), a widely adopted protocol providing data exchange in industrial environments. Particularly, we have worked to evaluate the possibility to perpetrate Slow DoS Attacks against ROS: While such kind of threats is recognized to be effective against conventional Internet services, its impact on industrial-related protocols is still unknown. By doing that, we designed a novel attack called SlowROS, aimed at causing a denial of service against a ROS service. We validated the attack by running tests on a real network composed of multiple components. While doing that, we found a potential vulnerability affecting the ROS software, allowing an attacker to exploit extremely long timeouts. By leveraging it during our tests, results show that the attack is successful and it is able to lead a denial of service against a ROS service, also impacting the functioning of the connected components.

In particular, while our tests consider a number of simultaneous connections N, which is much greater than the maximum number N_s accepted by the server, an extension of the current work may adopt lower values of N, approaching N_s. Other optimizations of the attack may consider different strategies for data sending, connections creation, or to minimize the attack bandwidth, by exploiting the potentially endless timeout adopted by the server. In addition, as we focused on ROS 1, further work may consider the use of ROS 2, although its development plan still misses relevant key points [26].

Considering the protection from the proposed threat, although identifying a single attacker may be trivial, the efficient detection of a distributed threat is challenging, since exchanged data is compliant to the protocol. Although we did not focus on mitigation strategies in this paper, in favor of the validation of the proposed thread, a preliminary analysis showed that a time-based approach, relying on connection tracking and forced connection shutdown upon expiration of an idle timeout might be viable. More specifically, on a Linux-based system, the Iptables `conntrack` module [33] can be used to track incoming connections to the ROS Master, so that its hash table eviction timer approximates an idle connection timer. Then, when `conntrack` evicts a connection a suitable tool, like `ss`, can be used to forcibly close it without altering the ROS Master code in any way. Future work on the topic may focus on perfecting such a technique and, in particular, find the optimal timeout value that maximizes the likelihood of telling apart legitimate (but slow) clients from malicious agents.

Acknowledgements This work was partially supported by project SERICS (PE00000014) under the MUR National Recovery and Resilience Plan funded by the European Union—NextGenerationEU.

References

1. Abeykoon I, Feng X (2019) Challenges in ROS forensics. In: 2019 IEEE SmartWorld, ubiquitous intelligence & computing, advanced & trusted computing, scalable computing & communications, cloud & big data computing, internet of people and smart city innovation (SmartWorld/SCALCOM/UIC/ATC/CBDCom/IOP/SCI), pp 1677–1682. https://doi.org/10.1109/SmartWorld-UIC-ATC-SCALCOM-IOP-SCI.2019.00299
2. America's cyber defense agency: ICS ADVISORY—multiple data distribution service (DDS) implementations (Update A) (2022). https://www.cisa.gov/news-events/ics-advisories/icsa-21-315-02. Accessed 24 May 2024
3. Antunes RA, Dalmazo BL, Drews PLJ (2022) Detecting data injection attacks in ROS systems using machine learning. In: 2022 Latin American robotics symposium (LARS), 2022 Brazilian symposium on robotics (SBR), and 2022 workshop on robotics in education (WRE), pp. 1–6. https://doi.org/10.1109/LARS/SBR/WRE56824.2022.9995917
4. Basan E, Sushkin N, Khabarov O, Makarevich O, Azarov I (2021) Analysis of attacks on robotic operation system. In: Proceedings of the international workshop on advanced in information security management and applications (AISMA 2021). CEUR Workshop Proceedings, pp 41–48
5. Botta A, Rotbei S, Zinno S, Ventre G (2023) Cyber security of robots: a comprehensive survey. Intell Syst Appl, 200237. https://doi.org/10.1016/j.iswa.2023.200237
6. Cambiaso E, Chiola G, Aiello M (2019) Introducing the slowdrop attack. Comput Netw 150:234–249. https://doi.org/10.1016/j.comnet.2019.01.007
7. Cambiaso E, Papaleo G, Chiola G, Aiello M (2013) Slow DoS attacks: definition and categorisation. Int J Trust Manag Comput Commun 1(3–4):300–319. https://doi.org/10.1504/IJTMCC.2013.056440
8. Caviglione L, Gaggero M (2020) Multiobjective placement for secure and dependable smart industrial environments. IEEE Trans Ind Inform 17(2):1298–1306. https://doi.org/10.1109/TII.2020.2978771
9. Değirmenci E, Kirca YS, Yolaçan EN, Yazici A (2023) An analysis of DoS attack on robot operating system. Gazi Univ J Sci 36(3):1050–1069. https://doi.org/10.35378/gujs.976496
10. DeMarinis N, Tellex S, Kemerlis VP, Konidaris G, Fonseca R (2019) Scanning the internet for ROS: a view of security in robotics research. In: 2019 International conference on robotics and automation (ICRA), pp 8514–8521. https://doi.org/10.1109/ICRA.2019.8794451
11. Dieber B, White R, Taurer S, Breiling B, Caiazza G, Christensen H, Cortesi A (2020) Penetration testing ROS. In: Koubaa A (ed) Robot operating system (ROS): the complete reference, vol 4. Springer International Publishing, Cham, pp 183–225. https://doi.org/10.1007/978-3-030-20190-6_8
12. Fuladi R, Baykas T, Anarim E (2024) The use of statistical features for low-rate denial-of-service attack detection. Ann Telecommun, 1–13. https://doi.org/10.1007/s12243-024-01027-3
13. Gogoi B, Ahmed T (2022) HTTP low and slow DoS attack detection using LSTM based deep learning. In: 2022 IEEE 19th India council international conference (INDICON). IEEE, pp 1–6. https://doi.org/10.1109/INDICON56171.2022.10039772
14. Inchara S, Keerthana D, Babu K, Mabel JP (2023) Detection and mitigation of slow DoS attacks using machine learning. In: AIP conference proceedings, vol 2917. AIP Publishing. https://doi.org/10.1063/5.0175586

15. Jeong SY, Choi IJ, Kim YJ, Shin YM, Han JH, Jung GH, Kim KG (2017) A study on ROS vulnerabilities and countermeasure. In: Proceedings of the companion of the 2017 ACM/IEEE international conference on human-robot interaction. HRI '17, Association for Computing Machinery, New York, NY, USA, pp 147–148. https://doi.org/10.1145/3029798.3038437
16. Liu Y, Al-Masri E (2023) Slow subscribers: a novel IoT-MQTT based denial of service attack. Clust Comput 26(6):3973–3984. https://doi.org/10.1007/s10586-022-03788-9
17. Malavolta I, Lewis G, Schmerl B, Lago P, Garlan D (2020) How do you architect your robots? State of the practice and guidelines for ROS-based systems. In: 2020 IEEE/ACM 42nd international conference on software engineering: software engineering in practice (ICSE-SEIP), pp 31–40. https://doi.org/10.1145/3377813.3381358
18. Mayoral-Vilches V, Carbajo UA, Gil-Uriarte E (2020) Industrial robot ransomware: Akerbeltz. In: 2020 fourth IEEE international conference on robotic computing (IRC). IEEE, pp 432–435. https://doi.org/10.1109/IRC.2020.00080
19. Mazurczyk W, Szary P, Wendzel S, Caviglione L (2019) Towards reversible storage network covert channels. In: Proceedings of the 14th international conference on availability, reliability and security, pp 1–8. https://doi.org/10.1145/3339252.3341493
20. The MITRE Corporation: ROS—Ros-comm—Security Vulnerabilities, CVEs, https://www.cvedetails.com/vulnerability-list/vendor_id-22167/product_id-73864/ROS-Ros-comm.html. Accessed 24 May 2024
21. Mokhamed T, Dakalbab FM, Abbas S, Talib MA (2022) Security in robot operating systems (ROS): analytical review study. In: The 3rd international conference on distributed sensing and intelligent systems (ICDSIS 2022), pp 79–94. https://doi.org/10.1049/icp.2022.2422
22. Mukhandi M, Portugal D, Pereira S, Couceiro MS (2019) A novel solution for securing robot communications based on the MQTT protocol and ROS. In: 2019 IEEE/SICE international symposium on system integration (SII), pp 608–613. https://doi.org/10.1109/SII.2019.8700390
23. Nugraha B, Murthy RN (2020) Deep learning-based slow DDoS attack detection in SDN-based networks. In: 2020 IEEE conference on network function virtualization and software defined networks (NFV-SDN). IEEE, pp 51–56. https://doi.org/10.1109/NFV-SDN50289.2020.9289894
24. Object Management Group, Inc.: DDS Security, version 1.1 (2018). https://www.omg.org/spec/DDS-SECURITY/1.1/About-DDS-SECURITY. Accessed 24 May 2024
25. Open Source Robotics Foundation, Inc.: ROS.org documentation. http://wiki.ros.org/Documentation. Accessed 24 May 2024
26. Open Source Robotics Foundation, Inc.: ROS 2 documentation: rolling documentation—features status (2024). https://docs.ros.org/en/rolling/The-ROS2-Project/Features.html. Accessed 24 May 2024
27. Pascoal TA, Fonseca IE, Nigam V (2020) Slow denial-of-service attacks on software defined networks. Comput Netw 173:107223. https://doi.org/10.1016/j.comnet.2020.107223
28. Qiu Q, Zhang X, Yao S, Chen Y, Chen G, Hua B, Ji J (2023) Training a non-cooperator to identify vulnerabilities and improve robustness for robot navigation. IEEE Robot Autom Lett 8(8):4465–4472. https://doi.org/10.1109/LRA.2023.3284379
29. ROS Industrial Consortium: Consortium home page. https://rosindustrial.org. Accessed 24 May 2024
30. van Rossum G (2024) The Python development team: the Python Library Reference—Release 3.12.3. Python Software Foundation
31. Teixeira RR, Maurell IP, Drews PL (2020) Security on ROS: analyzing and exploiting vulnerabilities of ROS-based systems. In: 2020 Latin American robotics symposium (LARS), 2020 Brazilian symposium on robotics (SBR) and 2020 workshop on robotics in education (WRE), pp 1–6. https://doi.org/10.1109/LARS/SBR/WRE51543.2020.9307107
32. Thapliyal N, Dhariwal S (2023) Protecting a nuclear power plant against a Stuxnet attack: power of computer security. In: 2023 international conference on communication, security and artificial intelligence (ICCSAI). IEEE, pp 598–603. https://doi.org/10.1109/ICCSAI59793.2023.10421224

33. The Netfilter Core Team: The netfilter.org "iptables" project (1999–2021). https://www.netfilter.org/projects/iptables/index.html. Accessed 20 Jun 2024
34. Trend Micro Research, TXOne, ADLINK, Alias Robotics, and ZDI (2022) Defending the supply chain: why the DDS protocol is critical in industrial and software systems. https://www.trendmicro.com/en_us/research/22/a/defending-the-supply-chain-why-dds-is-critical-in-industrial-and-software-systems.html. Accessed 24 May 2024
35. Tripathi N, Hubballi N (2018) Slow rate denial of service attacks against HTTP/2 and detection. Comput Secur 72:255–272. https://doi.org/10.1016/j.cose.2017.09.009
36. Vaccari I, Aiello M, Cambiaso E (2020) SlowITe, a novel denial of service attack affecting MQTT. Sensors 20(10). https://doi.org/10.3390/s20102932
37. Vaccari I, Aiello M, Cambiaso E (2020) SlowTT: a slow denial of service against IoT networks. Information 11(9):452. https://doi.org/10.3390/info11090452
38. Widodo A, Rochim AF, Warsito B (2023) Exploring low rate Dos attack detection methods: a bibliometric analysis and roadmap for further research. In: 2023 IEEE international conference on data and software engineering (ICoDSE). IEEE, pp 120–125. https://doi.org/10.1109/ICoDSE59534.2023.10291240
39. Winer D (2024) XML-RPC specification. UserLand Software. https://xmlrpc.com/spec.md. Accessed 24 May 2024
40. Wyrobek KA, Berger EH, Van der Loos HM, Salisbury JK (2008) Towards a personal robotics development platform: rationale and design of an intrinsically safe personal robot. In: 2008 IEEE international conference on robotics and automation, pp 2165–2170. https://doi.org/10.1109/ROBOT.2008.4543527
41. Yassein MB, Shatnawi MQ, Aljwarneh S, Al-Hatmi R (2017) Internet of things: survey and open issues of MQTT protocol. In: 2017 international conference on engineering & MIS (ICEMIS), pp 1–6. https://doi.org/10.1109/ICEMIS.2017.8273112
42. Zaim HÇ, Yolaçan EN, Yavanoğlu U (2023) Intelligent attack detection in ROS-based systems. In: 2023 IEEE international conference on big data (BigData), pp 5946–5950. https://doi.org/10.1109/BigData59044.2023.10386583

Building Type-Safe and Scalable Backends for Modern SaaS with ElysiaJS

S. Yeganathan and M. Nayan Thara

Abstract With the mounting complexities of modern backend systems, there is a growing need for technologies that deliver robustness, scalability, and enhanced developer experience (DX). This research paper explores a novel backend technology, constructed on the Bun runtime, that exhibits exceptional capabilities in catering to these demands: ElysiaJS. Through a comparative analysis involving varied stress levels, the performance of this emerging technology is scrutinized against other frameworks, using critical metrics such as RPS (requests per second) and average latency. Results exhibit the framework's unparalleled ability to handle increasing loads and maintain minimal latency. The paper further delves into its unique attributes, architectural patterns, and implementations through coding examples. Emphasizing end-to-end type safety intrinsic to an Eden Treaty and its adeptness to develop decoupled modules and employ dependency injection, the analysis posits this technology as a significant milestone in the landscape of backend system development. Therefore, the utilization of ElysiaJS presents a clear path towards crafting reliable, efficient, and maintainable software architectures.

Keywords Backend systems · Developer experience (dx) · ElysiaJS · Type safety · Scalability · High-performance · Bun runtime · Microservices

1 Introduction

Web applications have undergone significant evolution, progressing from basic static content to dynamic and feature-rich systems. Central to these advancements are the backend technologies supporting these applications. Traditionally, languages such as Python, Java, and Node.js have been favored for backend development. However, with the growing complexity of data and applications, developers are exploring alternatives promising enhanced efficiency, scalability, and crucially, type safety. Type

S. Yeganathan (✉) · M. Nayan Thara
Amrita School of Computing, Amrita Vishwa Vidyapeetham, Kollam, India
e-mail: research@yeganathan.me

© The Author(s), under exclusive license to Springer Nature Switzerland AG 2025

S. Femmam and P. Lorenz (eds.), *Recent Advances in Communication Networks and Embedded Systems*, Lecture Notes on Data Engineering and Communications Technologies 256, https://doi.org/10.1007/978-3-031-92449-1_10

safety, a fundamental concept in programming, plays a pivotal role in enhancing the performance and scalability of backend systems [1]. By enforcing strict type rules, errors related to data types are captured at compile time rather than runtime, resulting in more robust and predictable code. This proactive approach minimizes the likelihood of runtime errors, thus reducing debugging and maintenance efforts. Furthermore, type-safe languages often facilitate smoother code refactoring and scalability, empowering developers to make changes confidently with fewer concerns about type-related issues.

The quest for superior solutions has led to the emergence of a new generation of backend technologies aiming to redefine software architecture standards. Among these innovative solutions, ElysiaJS stands out as a promising newcomer. Built on the Bun runtime, it aims to elevate web application backends by combining peak performance with a strong emphasis on type safety and developer experience-factors often overlooked in conventional languages. By significantly mitigating type-related pitfalls at runtime, the framework reinforces the robustness of software systems. This reduction in debugging and maintenance overhead not only streamlines the software life cycle but also optimizes resource allocation across projects [2, 3]. The software development landscape is constantly evolving, and new technologies like ElysiaJS promise to revolutionize how we build modern SaaS backends.

This research contributes to the understanding of the framework's capabilities and benefits, exploring its potential to fundamentally alter web application development practices. We analyze its impact on the development process and discuss its broader implications for industry standards and best practices. Ultimately, this paper aims to add to the ongoing dialogue on backend development and inspire innovative approaches to building robust, scalable, and type-safe web applications.

2 Related Works

The domain of backend development has seen significant advancements aimed at improving performance, scalability, and developer productivity. Historically, frameworks such as Node.js and Express.js have dominated the landscape due to their non-blocking I/O and middleware architecture, as extensively documented by Tilkov and Vinoski [14]. These frameworks laid the groundwork for modern JavaScript-based backend development. However, the evolving complexity of web applications has necessitated the search for more efficient and type-safe solutions. The advent of TypeScript, as explored by Bierman et al. [15], marked a pivotal shift towards incorporating static types in JavaScript, greatly enhancing code reliability and maintainability.

Recent trends in backend systems emphasize the importance of modular architectures and advanced runtime environments. Frameworks like NestJS have integrated TypeScript extensively to offer a robust dependency injection system and modular design, as detailed by Schwark et al. [16]. Concurrently, the Bun runtime has emerged as a promising alternative to Node.js, offering superior performance benchmarks, as evidenced by Reiser and Gienger [17]. Our research specifically evaluates ElysiaJS, a framework built on the Bun runtime, focusing on its performance, type safety, and developer experience. Through comprehensive benchmarking and architectural experimentation, we found ElysiaJS to excel in handling high loads with minimal latency, making it a compelling choice for modern backend systems. This study builds on the existing literature by providing empirical evidence of ElysiaJS's capabilities and proposing optimal architectural patterns to leverage its strengths.

3 Performance Analysis

For a comprehensive evaluation, we adhered to a structured benchmarking process to ensure a fair comparison of each framework's capabilities. We benchmarked the latest versions of popular JavaScript frameworks: Elysia 1.0, Express 4.19, Hono 4.2, Koa 2.15, and NextJS-Express 10.3. These frameworks were chosen for their widespread use and reputation in building software backend systems, ensuring that our selection reflects a diverse and representative sample of commonly used frameworks in the industry, thereby minimizing potential biases. Using the latest versions ensures our evaluation reflects current optimizations, which might have impacted the results as newer versions often include improvements and bug fixes. The test was conducted on 2024-04-08 (d24b381). We used the open source benchmarker tool for the analysis [4]. Delving into the test setup, an initial step involved establishing a test environment for each examination within isolated Docker containers. Table 1 shows the hardware setup used for the test.

The benchmarking process employs Ruby-based tools, including wrk for collecting results, PostgreSQL for storing data, and jq for processing Docker metadata. The frameworks are benchmarked using wrk (threads: 8, timeout: 8, duration: 15 s) with 64, 256, and 512 concurrency, assessing how each framework responds to different stress levels. Subsequently, load generation is initiated [5] to produce HTTP requests, mimicking real-life network traffic scenarios.

This meticulous approach enables the measurement of the scalability and efficiency of the frameworks under varying loads, providing valuable insights for comparative analysis, with a focus on key performance metrics: total requests per second (RPS) and average latency.

Table 1 Hardware used for the benchmark

Component	Specification
CPU	8 Cores (AMD FX-8320E Eight-Core Processor)
RAM	16 GB
OS	Linux

While total requests per second and average latency are informative metrics, this study focuses on these two to keep the analysis concise and manageable. We recognize that real-life applicability may vary and that other performance metrics could offer a more comprehensive evaluation of the frameworks' capabilities. However, by concentrating on RPS and latency, we aim to provide an initial understanding of the frameworks' potential.

We acknowledge that external factors such as network conditions and the complexity of the test cases could have influenced the results, despite our efforts to control these variables through isolated Docker environments.

3.1 Total Requests Per Second Metrics

This metric measures a framework's throughput by quantifying the number of requests it can handle per second. A higher RPS indicates better scalability and efficiency, enabling the framework to manage increasing loads without compromising performance[5].

Figure 1 illustrates the performance of different frameworks under load, with the x-axis representing concurrency (simulated multiple users in a load test) and the y-axis depicting the number of requests processed per second. ElysiaJS demonstrates the highest RPS across all concurrency levels, peaking at approximately 140,000 RPS with a concurrency level of 512. Hono follows as the second-best performer, achieving a maximum of approximately 70,000 RPS at 256 concurrency, while Koa closely trails with approximately 55,000 RPS at 512 concurrency. Express and NextJS-Express exhibit lower RPS values, further highlighting ElysiaJS's superior performance.

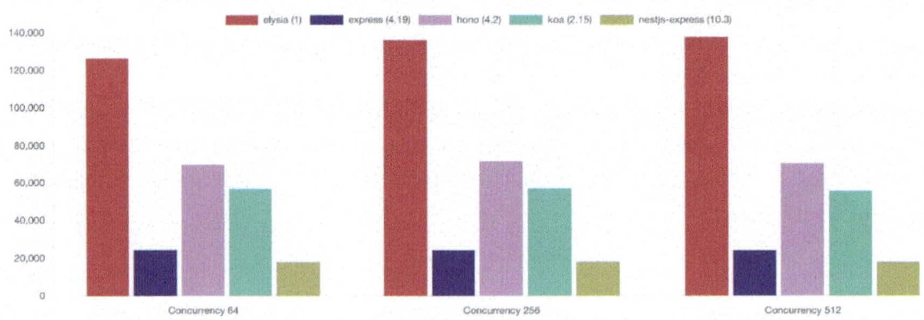

Fig. 1 Total requests per second

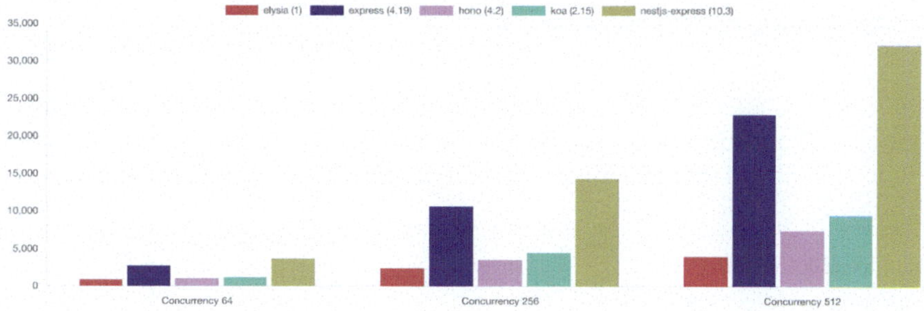

Fig. 2 Average latency (ms)

3.2 Latency Metrics

Latency, denoting the time taken for a request to be processed by the backend framework, is crucial for ensuring swift response times and enhancing user experience [5].

Figure 2 presents the average latency of different frameworks at various concurrency levels. In the graph, the x-axis represents concurrency (number of concurrent requests) and the y-axis represents average latency in milliseconds (ms). ElysiaJS consistently maintains the lowest latency across all levels, with a minimum of 5000 milliseconds approx at 512 concurrency. Hono closely follows with 6000 milliseconds approx at 512 concurrency, while Koa ranks third with 7000 milliseconds at 512 concurrency. Express and NextJS-Express exhibit higher latency values compared to ElysiaJS.

Thus, our analysis underscores ElysiaJS's exceptional performance in terms of both RPS and latency, positioning it as an attractive choice for developers seeking high-performance and scalable backend solutions. The results provide an initial understanding of the frameworks' potential, but their generalizability to real-world scenarios may vary due to differences in hardware setups and benchmarking tools. Further testing in varied environments would be needed for a more thorough evaluation.

4 Methodology

For this study, we harness the power of ElysiaJS, focusing particularly on its architectural patterns and type-safety suite that have applicability in the construction of backend systems. It's noteworthy to add that the framework typically utilizes Dependency Injection (DI) for sharing the types with the instances. However, it also presents us with the flexibility to incorporate a variety of different paradigms along with it.

4.1 Architectural Paradigms

Architectural paradigms serve as guiding principles in structuring systems. They dictate the overall organizational pattern of the system's software components, defining their interactions and responsibilities [6]. In this context, two pivotal paradigms warrant further exploration:

- Composable Modules
- Dependency Injection.

The effectiveness of these paradigms, unearthed through practical application, delivers valuable insights for harnessing their power in real-world scenarios.

4.1.1 Composable Modules

Composable modules in a software backend system refer to independent, self-contained components that perform specific functions in isolation. This structure enables individual modules to operate and evolve independently without directly impacting others. Decoupling enhances system flexibility, streamlines updates, and increases resilience, facilitating a more robust and efficient software architecture [7, 8].

The core idea is to abstract code logic into independent, logically composable units. These units shouldn't have cyclical dependencies with other modules and should function autonomously. By doing this, we can interchange modules based on functional requirements, enhancing the "plug-and-play" nature of these units. To maintain this system, these modules shouldn't include specific validations, as it limits their reusability and complicates maintainability. This pattern is employed in Elysia, where we perform validation at the controller/route level while using composable units for additional logic as needed. With such decoupled and interchangeable units, the "Controller Per Entity" design pattern becomes essential in such systems. Each controller embodies the business logic related to its respective entity, such as auth, project, team, etc.

Listing 1.1 illustrates a composable module named *getAuthUser*, which is designed to fetch an authenticated user from the database based on provided options such as email or userId. This module encapsulates reusable logic that can be utilized across different parts of the system.

In Listing 1.2, the *getAuthUser* module is employed within the *authRoutes* controller, which handles authentication-related routes. Specifically, during the signup process, it fetches the authentication user data using the email provided in the request body.

Listing 1.3 demonstrates another instance where the *getAuthUser* module is reused in a different route defined in *someRoutes*. Here, the module is called to retrieve user information based on the userId available in the session, showcasing the module's reusability across different contexts.

```
// Module parameter types
type Options = {
    email?: string
    userId?: string
}

/**
 * Composable module that fetches a auth user
 * @param opts - email or userId
 * @returns auth user object from the database
 */
export async function getAuthUser(opts: Options) {
    // write you abstracted re-usuable logic here ...
    // finally execute db call if needed
    return await db.selectFrom('auth.users').selectAll().executeTakeFirst()
}
```

Listing 1.1 getAuthUser module

```
// authRoutes.ts
// This controller contains all the auth related routes

const authRoutes = new Elysia({ name: 'routes:auth' })
.post('/auth/signup', async ({ body: data }) => {
    // getAuthUser module is used to fetch auth user data
    let user = await getAuthUser({ email: data.email })
    // more code logic goes here...
}
```

Listing 1.2 Controller where getAuthUser module is used

```
// someRoutes.ts
const someRoutes = new Elysia({ name: 'routes:some' })
.use(apiAuthenticate) // middleware that adds session & user to the ctx
.get('/me', async ({ user }) => {
    // module is reused here...
    return await getAuthUser({ userId: user?.id })
})
```

Listing 1.3 Another route where the module is used

Key features of the approach:

1. Unit-Testing Ease: This approach prioritizes unit testing for each module to guarantee consistently expected results.
2. Framework Agnostic: By focusing on the individual modules' independence, we avoid over-reliance on a specific framework and its features. This versatility

facilitates easier migration to new practices or approaches in the ecosystem without dramatic changes or breakdowns.

3. Great Maintainability: Each module can have individual JSdoc, enabling efficient iteration based on versions. For example, **fetch_projects** might be a module that retrieves team-specific projects. If we need to optimize certain parts, we can simply create an improved version with a clearer name, like **fetch_projects_v2**. This facilitates smooth optimizations and migrations.
4. Type Safety: Within each module, we ensure type safety for both arguments and returned values, thereby greatly enhancing overall type security.

Setbacks: While a composable approach offers numerous benefits and features, it also has certain drawbacks. Writing code can sometimes pose challenges in extracting logic in a generic manner. Developers often find themselves duplicating code across the codebase, complicating scalability and maintenance [9]. These issues can be mitigated with disciplined coding conventions and thorough documentation.

4.1.2 Dependency Injection (DI)

Dependency Injection (DI) is a technique in software backend systems where one object supplies the dependencies of another object, promoting loose coupling and enhancing testability [10]. In particular, with frameworks like Elysia, it becomes seamless to accomplish DI. We can create a level of dependency injection (DI) that equals or even surpasses NestJS in some cases. Every module or plugin to be injected is defined within an Elysia instance. Each Elysia instance is considered a controller, housing any logic that can be decoupled and placed inside a service layer as a separate elysia instance, available for injection if needed. Essentially, the goal is to designate everything as an Elysia instance, utilizing them as required with other controller instances. This strategy results in enhanced type safety and autocompletion. Listing 1.4 demonstrates a comprehensive code sample of dependency injection in Elysia.

```
// someModel.ts
const nameSchema = t.Union([ t.Literal("John"), t.
    Literal("Doe")]);
export type Name = Static<nameSchema>

export const someModel = new Elysia({ name: "
    someModel" })
  .model({
    nameParams: t.Object({ name: nameSchema }),
  })

// someService.ts
export const someService = new Elysia({ name: "
    someService" })
  // use .use(someOtherService) to depency-inject
    some other service
  .derive(({ /**/ }) => {
```

```
    // define useful functions here, and return them as an object
    // Name is type from someModel.ts
    const someUsefulFunction = (name: Name) => {
      return "Hi, "+name;
    }
    return { someUsefulFunction };
  })

// someController.ts
export const someController = new Elysia({ prefix: "/something" })
  // depency-inject the model and service
  .use(someModel)
  .use(someService)
  .get("/:name", ({ params, someUsefulFunction }) => {
    // name is validated and typed as Name from someModel.ts
    const { name } = params;
    return someUsefulFunction(name);
  }, {
    // autocompleted schema name
    params: "nameParams"
  })
```

Listing 1.4 Simple dependency-injection code sample

Key features of the approach:

1. Minimal type gymnastics due to the tightly packed layers
2. Full utilization of the framework's potential, enhanced by improved type safety
3. Facilitation of MVC implementation with ease.

Setbacks: Setbacks: The framework adherence often increases complexity in our codebase if not managed carefully. Difficulty arises when attempting to migrate to a different framework or when needing to employ Elysia instances along with service logic outside the existing framework. We find ourselves reliant on this specific framework as a result. It is crucial to avoid cyclic dependencies between layers, as these can lead to challenging debugging when functionality issues occur.

4.2 *Type Safety on the Client-Side*

Eden Treaty in the ElysiaJS framework provides end-to-end type safety. It is an object representation for server-frontend interaction, designed to work smoothly in a frontend based on JavaScript frameworks. This allows for changes in the type on the server that instantly reflect on the client, aiding with auto-completion and enforcing type constraints. It guarantees type safety, automatic completion, and efficient error handling. The server type of Elysia is imported directly to the frontend to consume the Elysia API. Through Treaty, Elysia ensures type safety across the full stack of the app, paving the way for quick identification and elimination of minor bugs via validations. It is an easy and efficient method to achieve type safety in the frontend [12].

Listing 1.5 demonstrates the usage of Eden Treaty for ensuring type safety. In the server.ts file, an Elysia server is created and configured with a simple GET route that returns a greeting message. The type of the server application is then exported. In the client.ts file, the Treaty library is used to connect to the server. By importing the server type, the client can make requests with guaranteed type safety, ensuring that the response type is correctly handled and validated, as shown in the code snippet where the client makes a GET request to the /hi route.

```
// server.ts
// Create and configure the server
const app = new Elysia()
    .get('/hi', () => 'Hi Elysia!')
    .listen(3000);

// Define the App type
export type App = typeof app;

// client.ts
import { treaty } from '@elysiajs/eden';
import type { App } from './server';

// Connect to the server and make a request
const app = treaty<App>('localhost:3000');
const { data, error } = await app.hi.get();
// Output response type: 'Hi Elysia'
```

Listing 1.5 Eden Treaty Usage

5 Results and Discussion

At the heart of the Elysia framework lies a sophisticated system of lifecycle events, akin to a stack. These events enable the interception of crucial stages at predefined junctures, as illustrated in Fig. 3. Think of it as a roadmap for your server's journey, with waypoints where developers can step in and tailor the behavior to meet specific requirements. Elysia enhances the development process through its advanced feature set and robust type safety, promoting efficient and error-free development. It offers runtime type validation through the use of models in route hooks, either locally or globally, as shown in Listing 1.6. In scenarios involving multiple controllers with the same types, Elysia allows the creation of a model with a .model hook, which can then be injected into other instances as necessary. It employs type box for runtime type validation. It also provides several ready-to-use hooks such as the onError hook,

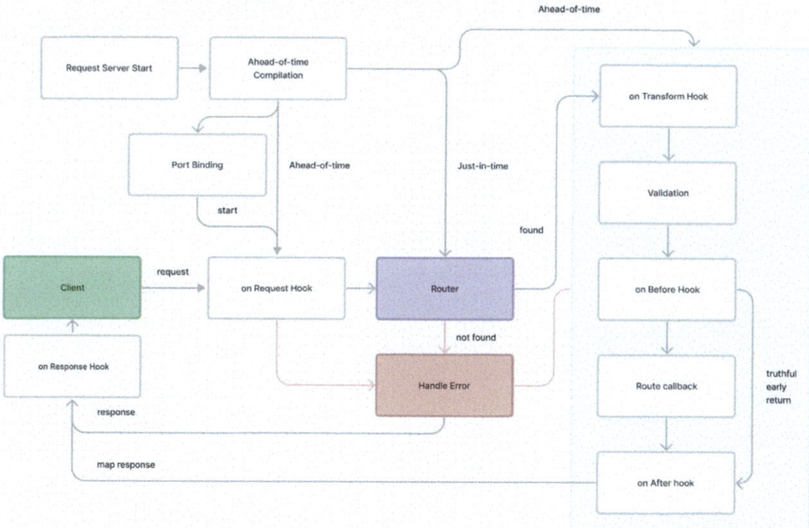

Fig. 3 Elysia's life cycle events (*Source* elysiajs.com)

enabling custom error handling, and easing the construction of an API server (see Listing 1.7).

Similar to other JS frameworks, It establishes a default context for each request, encapsulating crucial properties such as path, body, query, params, headers, request, store, cookie, set, and error. Developers can extend or add properties to this context with assured type safety. The extension of the context is achievable via several methods: state, decorate, resolve, and derive. The 'state' facilitates a shared mutable object across the Elysia app, excellent for introducing additional functions, singletons, or immutable properties to all handlers. The 'decorate' method is suitable for assigning additional functions or establishing an immutable property for all handlers, with the understanding that these values are read-only and never reassigned. This approach is particularly beneficial for defining aspects like a database client or logger and leveraging them inside controllers. However, decorate and state are relevant only when information is required during the server's boot-up stage. Derive and resolve methods offer the capability to add context when a request occurs. Unlike 'decorate' and 'state,' 'derive' can access the aforementioned request properties. Conversely, 'resolve' functions as a safer version of 'derive,' operating in the onBeforeHandle hook, while 'derive' functions afterward.

```
import { Elysia, t } from 'elysia'

new Elysia()
    .post('/', () => 'Hello World!', {
        body: t.Object({
            name: t.String()
        })
```

```
    })
    .listen(3000)
```

Listing 1.6 Local hook types declaration

```
// Define SignDTO model in same or different file
   eg. models.ts
const SignDTO = t.Object({
    username: t.String(),
    password: t.String()
})

new Elysia()
    .post('/sign-in', ({ body }) => body, {
        body: SignDTO,
        response: SignDTO
    })
```

Listing 1.7 Reference model

Plugins in Elysia are essentially standalone instances utilized as decoupled modules, offering convenience and flexibility. Named plugins ensure deduplication, thereby preventing redundant registrations. Elysia features an assortment of pre-built common plugins such as 'Bearer', 'CORS', 'JWT', 'HTML', 'Static', and 'Cron', alongside numerous community plugins catering to authentication support, file system routes, and more. Eden treaty combined with bun's test suite is perfect for API testing, confirming that an API endpoint returns the correct response to a given request. CI/CD pipelines can be used to run these tests even before deploying them to production. This pattern can be leveraged to create a unit test with end-to-end type safety and type-level tests all at once.

With the significant findings from our study, we have opted to use Elysia as the backend JavaScript framework for our project GitWonk, an open-source technical documentation tool built from the ground up for modern teams. It fosters effortless cooperation, allowing various types of docs including product details, in-house know-how, and APIs to be published. The project development process emphasizes the necessity for module design and rigorous testing, spanning both unit and API levels. Plans are underway to boost frontend type safety using the Eden Treaty, alongside advanced caching.

6 Conclusion

In conclusion, this research unequivocally establishes ElysiaJS as a standout choice in backend technology, distinguished by its exceptional ability to manage high loads with minimal latency. Our thorough examination highlights numerous aspects of the framework that contribute to its superiority in backend system development. One standout feature is its commitment to end-to-end type safety, exemplified by the implementation of the eden treaty. This foundational element ensures a robust codebase, reducing the occurrence of runtime errors and bolstering system reliability. It offers flexible architectural patterns such as composable modules and dependency injection. These features not only facilitate scalable software architectures but also enhance maintainability over time. The framework's design principles foster an environment where developers can build robust applications efficiently.

As for future work, further exploration could focus on optimizing ElysiaJS's performance even further and exploring innovative applications to maximize its potential in diverse use cases. By continuing to explore and refine its capabilities, ElysiaJS stands poised to remain a formidable choice for developers seeking robust, high-performance backend solutions in modern software development.

References

1. Jansen Remo H, Vilic V, De Wolff IG (2016) Modern JavaScript development. Packt Publishing Ltd., TypeScript
2. Islam MS (2023) JavaScript alternative (TypeScript) and its effectiveness in web development
3. Bogner J, Merkel M (2022) To type or not to type? A systematic comparison of the software quality of JavaScript and typescript applications on GitHub. In: Proceedings of the 19th international conference on mining software repositories, pp 658–669
4. Benchmark-web-frameworks source code. GitHub repository. https://github.com/the-benchmarker/web-frameworks
5. Kadi S (2021) Measuring maintainability and latency of node. JS frameworks
6. Shaw M, Garlan D (1996) Software architecture: perspectives on an emerging discipline. Prentice-Hall, Inc
7. Raghavan B, Casado M, Koponen T, Ratnasamy S, Ghodsi A, Shenker S (2012) Software-defined internet architecture: decoupling architecture from infrastructure. In: Proceedings of the 11th ACM workshop on hot topics in networks, pp 43–48
8. Bjuhr O, Segeljakt K, Addibpour M, Heiser F, Lagerström R (2017) Software architecture decoupling at Ericsson. In: 2017 IEEE international conference on software architecture workshops (ICSAW). IEEE, pp 259–262
9. Mo R, Cai Y, Kazman R, Xiao L, Feng Q (2016) Decoupling level: a new metric for architectural maintenance complexity. In: Proceedings of the 38th international conference on software engineering, pp 499–510
10. Seemann M, van Deursen S (2019) Dependency injection principles, practices, and patterns. Simon and Schuster
11. Nasehi SM, Maurer F (2010) Unit tests as API usage examples. In: 2010 IEEE international conference on software maintenance. IEEE, pp 1–10
12. ElysiaJS website, https://elysiajs.com. Accessed 7 Apr 2024
13. Prasanthan A, Anand KS, Nair BP, Gautham Santhosh K, Swaminathan J (2023) Low code backend as a service platform. In: 2023 world conference on communication & computing (WCONF), RAIPUR, India, pp 1–8. https://doi.org/10.1109/WCONF58270.2023.10234969
14. Tilkov S, Vinoski S (2010) Node.js: using JavaScript to build high-performance network programs. IEEE Int Comput 14(6):80–83. https://doi.org/10.1109/MIC.2010.145
15. Bierman G, Abadi M, Torgersen M (2014) Understanding TypeScript. In: European conference on object-oriented programming (ECOOP), pp 257–281
16. Schwark T, Rickenberg M, Grinberg N (2018) TypeScript with NestJS: a scalable and modular approach to backend development. J Web Eng 17(5–6):475–495
17. Reiser R, Gienger F (2022) Bun: a new JavaScript runtime for high performance. J Softw Perform Eng 12(3):210–225

Identifying the SIM Card Authentication Algorithm Using Power Consumption Analysis

Anastasia Fedorovna Ponomareva, Alla Borisovna Levina, Efim Borisovich Ashmarov, Igor Alekseevich Zikratov, and Vladislav Evgenevich Trifonov

Abstract With the development of cellular communication, securing SIM cards becomes an increasingly relevant task. Side-channel attacks (SCA) are among the most promising attack vectors for embedded cryptographic devices, including SIM cards. This study focuses on vulnerabilities SIM cards and evaluates their resistance to side-channel attacks. We employed a black-box strategy and, through power consumption analysis, identified the authentication algorithm in the examined SIM cards. The results of the study confirm that the investigated SIM cards lack countermeasures against side-channel attacks and also demonstrate potential vulnerability of the SIM card authentication algorithm to power analysis differential attacks.

Keywords SIM · USIM · MILENAGE · SCA · Information security

1 Introduction

With the proliferation of mobile devices such as smartphones and smart gadgets, along with their integration into everyday life, SIM cards have become a key element in ensuring communication and security. The extensive use of mobile devices makes them an attractive target for potential attacks.

Side-channel cryptanalysis is a branch of cryptography in which confidential information is extracted from the physical implementation of the target cryptosystem. This distinguishes it from other forms of cryptanalysis, in which algorithms and

A. F. Ponomareva · A. B. Levina (✉) · E. B. Ashmarov · V. E. Trifonov
Saint Petersburg Electrotechnical University "LETI", St. Petersburg, Russia
e-mail: alla_levina@mail.ru

I. A. Zikratov
Saint Petersburg State University of Telecommunications, St. Petersburg, Russia

© The Author(s), under exclusive license to Springer Nature Switzerland AG 2025
S. Femmam and P. Lorenz (eds.), *Recent Advances in Communication Networks and Embedded Systems*, Lecture Notes on Data Engineering and Communications Technologies 256, https://doi.org/10.1007/978-3-031-92449-1_11

underlying computational tasks are attacked. There are many ways to leak information from electronic devices. The most striking examples of this are temperature, energy consumption, time spent on calculations, acoustics and electromagnetic radiation.

Devices running on embedded systems are usually more vulnerable to side-channel attacks such as Energy Consumption Analysis and Differential Energy Consumption Analysis. One of the earliest power analysis attacks was the Differential Power Analysis (DPA) attack on DES implementation in 1999 [1]. This attack resulted in the recovery of secrets from embedded cryptographic chips. Additionally, in 2009, a successful attack was conducted on an embedded device (chip), modeled as a black box, using Mutual Information Analysis (MIA) based on the analysis of differential power [2].

Energy Consumption Attacks (Power Analysis Attacks)—is another form of side-channel attacks based on analyzing the power consumption of a cryptographic device while it performs the encryption operation. The attacker can non-invasively extract secret keys and other information from the device by combining with other crypt-analysis techniques. An attacker analyses the device's energy consumption while performing a cryptographic operation, such as an analysis of the energy consumption at different stages of an algorithm. Devices running on embedded systems, such as smartphones and other mobile devices, are more vulnerable to these attacks [3]. Simple Power Analysis (SPA) and Differential Power Analysis (DPA) are two types of cryptographic attacks that exploit information derived from changes in the power consumption of a device during operation. These attacks can be used to extract sensitive information, such as passwords or encryption keys [4].

SPA (Simple Power Analysis) is a technique that analyzes the power consumption of a cryptographic device during its operation. During a SPA attack, an attacker monitors the device's power consumption and attempts to correlate changes in power usage with specific operations or data processing. By examining the patterns of power usage, an attacker may gain insight into the inner workings of the cryptographic algorithm. DPA (Differential Power Analysis) is an advanced type of SPA that concentrates on the statistical examination of differences in energy use between various operations or data inputs. Instead of examining absolute values of power use, DPA contrasts the energy signatures of different inputs or operations. Through statistical analysis of differences, an adversary may identify patterns that disclose confidential information, like secret cryptographic keys [1].

The purpose of this work is to develop a hardware and software complex to identify the authentication algorithm in the studied SIM cards using power consumption analysis. The objectives of the work are:

- To study existing authentication algorithms in mobile networks
- Design and implementation of a laboratory stand for removing power traces.
- To conduct tests during which it is possible to identify the authentication protocol in LTE SIM cards.

The MILENAGE algorithm is a set of mathematical functions based on the AES-128 standard, it is used for an authentication in smartphones. MILENAGE is used for authentication and key agreement in UMTS/LTE networks [5]. The considered algorithm was invulnerable to any known attacks until 2014. First successful attack was conducted on MILENAGE in 3G/4G USIM cards, which utilize the Advanced Encryption Standard (AES) Rijndael algorithm in 2015 [6]. In fact, with a DPA attack, an attacker could recover a subscriber's key and other secrets recorded by the mobile network operator from several hundred traces of unprotected USIM cards that were analyzed. Some SHA can be found in works [7–9].

The results of this work were based on power analysis differential attacks and demonstrated that encryption algorithms in 3G/4G USIM cards relied on unprotected AES implementations from side-channels.

This study will present the execution of SIM card analysis using the "black-box" strategy, secret parameter values, authentication algorithm, and measures to counteract side-channel attacks—all currently unknown.

The paper is organized as follows: Sect. 2 presents an Authentication and Key Agreement (AKA) protocol, which is implemented in SIM cards, Sect. 3 presents an attack on this protocol, Conclusion is presented in Sect. 4.

2 Authentication and Key Agreement (AKA) Protocol

3GPP [10] is a collaborative agreement established in 1998 among international telecommunications standardization organizations. Its goal, among others, is to develop technical specifications for third and fourth generation (3G/4G) mobile cellular networks. To enhance the security of the authentication procedure over these next-generation networks, 3GPP proposed the Authentication and Key Agreement (AKA) protocol [11]. The figure illustrates a simplified structure of the protocol in a unified form for 3G/4G networks (see Fig. 1).

User equipment consists of a SIM card and a smartphone, which are owned by the subscriber. The main components of the user equipment are the International Mobile Subscriber Identity (IMSI), a symmetric secret key (K), operator confidential parameters (OPc, $r_1 - r_5$, $c_1 - c_5$), and a cryptographic algorithm.

Serving networks act as relays for home networks, to which user equipment can connect. Home networks maintain a database of USIM cards and respond to authentication requests.

The protocol begins with an identity request, to which the SIM card responds with the IMSI value. Upon an authentication request, the Authentication Center (AuC) generates a random number (RAND), sets a sequence number (SQN), and computes the algorithm (a set of functions f_1–f_5)) to derive the authentication parameters. An authentication vector (AUTN), a 128-bit authentication token, is then generated and represented by the following formula 1:

$$AUTN = (SQN \oplus AK) \| AMF \| MAC \tag{1}$$

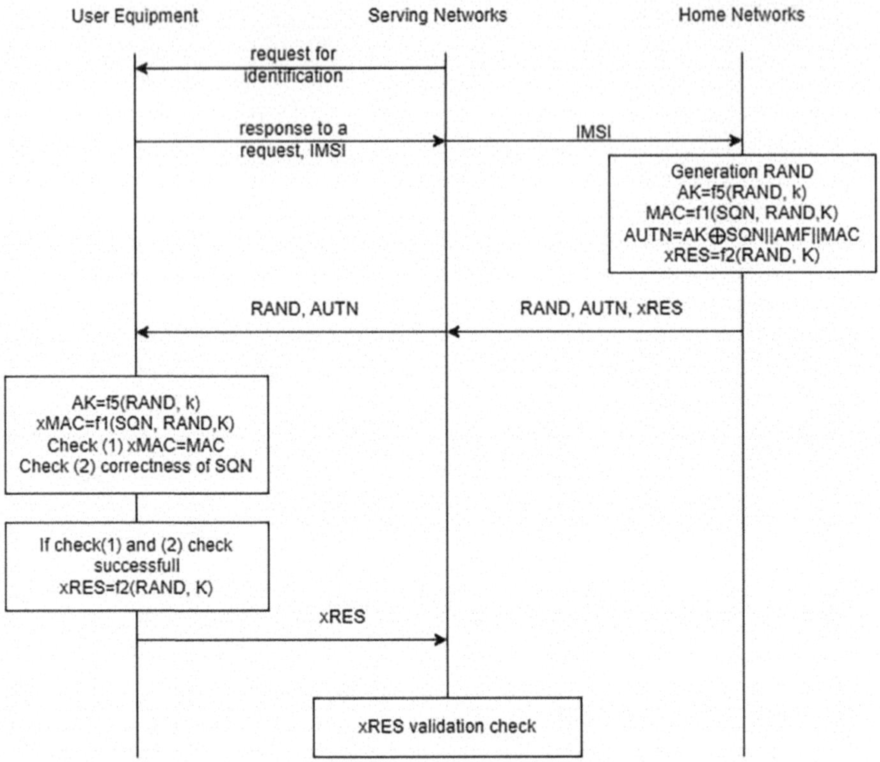

Fig. 1 AKA protocol

AK is a 48-bit anonymity key, AMF is a 16-bit authentication management field, and MAC is a 64-bit message authentication code.

The USIM card then receives the RAND and AUTN values and begins calculating the SQN and xMAC (expected MAC) values using functions f_5 and f_1, respectively. The f_5 function calculates the AK value to find the SQN as SQN = (SQN $\oplus$ AK) $\oplus$ AK. The USIM verifies the received values; if the SQN is out of range or the xMAC does not match the MAC, the authentication stops, and an appropriate error message is displayed. The other functions (f_2–f_4) will not be computed. If the check is successful, the remaining functions are computed to find the xRES (expected RES) value. If xRES matches the RES value, the authentication is considered successful.

The AKA protocol is compatible with MILEAGE algorithm. This article focuses on the MILENAGE algorithm, which is a set of mathematical functions f_1–f_5 based on AES-128 (see Fig. 2).

The Advanced Encryption Standard (AES [12]) is a symmetric block cipher used to protect data, working with blocks of 128 bits. Previously, the input data is divided into blocks of 16 bytes, if the full size is not a multiple of 16 bytes, then the data is expanded to a multiple of 16 bytes. The names of the operations used in AES are:

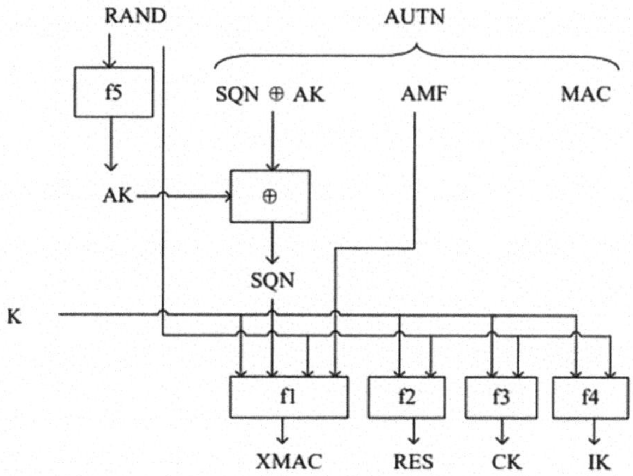

Fig. 2 MILENAGE algorithm [6]

state, addRoudKey, SubBytes, ShiftRows, and MixColumns. AES with a 128-bit key K uses 10 rounds (iterations).

The AES encryption algorithm (E_k) performs the following steps:

- The plaintext m (128-bit block) is converted using the state function, which divides the input text into 4 × 4 (128-bit) matrices.
- Next, the whitening function is performed. This function will perform the XOR operation of the source text with the key.
- The following KeyExpansion function calculates the round encryption key.
- The SubBytes, ShiftRows, MixColumns and addRoudKey operations are performed with the calculated round key.
- The last 2 steps are repeated 9 times with the corresponding round keys.
- In the last round, the SubBytes and ShiftRows functions are executed. Next, the XOR operation is performed with the received text.

Let's take a closer look at each of the transformations presented above:

- State—performs the XOR operation with the plaintext m and the encryption key K.
- KeyExpansion—calculates the round encryption key. The extended key consists of 44 four-byte words (w_i): 4 words per main key and 4 words per 10 round keys. Thus, the total length of the extended key is 1408 bits. The key expansion operation uses the Rcon array. To begin with, the four words of the main key are transferred to the first four words of the extended key. If the number i is divisible by 4 without remainder, then the value of w_i is calculated using the formula 2. Otherwise, w_i is calculated using the formula 3.

$$w_i = SubBytes(RotByte(w_{(i-1)})) \bigoplus Rcon_{(\frac{i}{4})} \tag{2}$$

$$w_i = w_{(i-4)} \bigoplus w_{(i-1)} \tag{3}$$

The RotByte operation cycles the byte of the source word: $\{x_0, x_1, x_2, x_3\} \rightarrow \{x_3, x_0, x_1, x_2\}$.

- SubBytes is the replacement of state bytes according to the S-box. Each byte is represented as two hexadecimal numbers b = (x, y), where x is defined by the 4 highest digits of b, and y by the 4 lowest. The 16 × 16 S-box table contains values to replace the original byte: the value b' at the intersection of row x and column y of the S-box is used as a replacement for the original byte b.
- ShiftRows—cyclic shift of state strings. The zero line remains in place, the first one is shifted to the left by 1 byte, the second by 2 bytes and the third by 3, respectively.
- MixColumns—multiplication of each state column by a fixed matrix. Thus, a linear transformation is performed over the state columns. Moreover, multiplication and addition are performed according to the rules described above.
- AddRoundKey—the round key is incrementally added to the state using a bitwise XOR.

In next Section we will illustrate the attack which helps to analyze SIM card authentication algorithm.

3 Implementation of the Attack

Since for the purposes of this work it is necessary to determine the authentication algorithm in the studied LTE SIM cards, for this it is necessary to perform the first three stages of the vulnerability detection methodology, as well as perform a SPA to determine the encryption algorithm.

LTE (Long-Term Evolution) is a standard for wireless high-speed data transmission used by mobile phones and other devices. It was developed by the 3GPP, a consortium that creates specifications for mobile communication, and is defined in a series of documents called Release 8, with minor updates described in Release 9 [5].

To implement the attack on energy consumption, a dual-interface card reader ACR1281U-C1 was used in the experiment, providing access to both contact and contactless smart cards in accordance with ISO 7816 [13] and ISO 14443 [14] standards. A shunt resistor 75SHCM-M3 100 A CL. 0.5 was used to obtain power, as well as a two-channel digital oscilloscope of the PicoScope 5000 series AKIP-75244V (see Fig. 3).

The SIM card exchanges data via the Application Protocol Data Unit (APDU) protocol. APDU is a data exchange format between the card and the terminal. The terminal sends an APDU command (C-APDU), and the card issues an APDU response (R-APDU). The response format is given in Table 1.

Each header element is stored on one byte and is mandatory. The header elements are as follows:

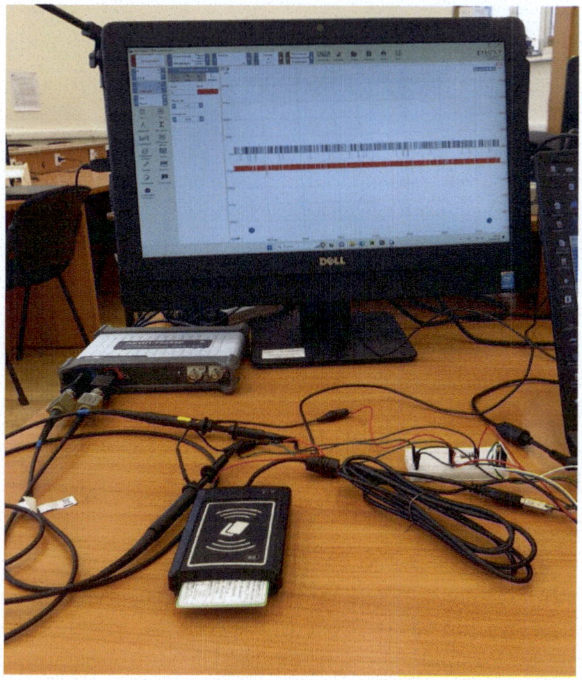

Fig. 3 Implementation of the stand

Table 1 The C-APDU format

Heading	Body
CLA INS P1 P2	[Lc] [Data] [Le]

- CLA (Class byte) contains the command metadata;
- INS (Instruction byte) contains the instruction code;
- P1 and P2 are the first and second parameters of the command.

The body elements are as follows:

- Lc is responsible for the length of the Data element in bytes;
- Data is the data of the command;
- Le is the expected length of the response data in bytes.

Table 2 shows the format of the APDU response.

The body is present when commands that expect to receive data from the card are executed successfully.

Table 2 The R-APDU format

Heading	Body
CLA INS P1 P2	[Lc] [Data] [Le]

Fig. 4 APDU commands

To execute the attack, it's necessary to initiate the internal authentication algorithm on the targeted SIM card. To invoke the authentication algorithm, first, the directory file EF is opened to access the EF file directory, followed by launching the USIM application. Then, the EF file for sending IMSI is invoked. Finally, the INTERNAL AUTHENTICATE function is called with random values RAND and AUTN. The terminal responds with "98 62"—security management: authentication error, incorrect MAC (see Fig. 4).

To perform the "measurement" step, Piton code was developed to efficiently collect traces. The main task of this software is to record traces from the oscilloscope during the execution of the authentication protocol in a.csv file, as well as record the input authentication parameters.

A low-pass filter was used for the "signal processing" stage in the experiment. This is necessary to smooth out irregularities and eliminate small fluctuations [8]. When reading the signal, the formula 4 must be fulfilled:

$$pot = (1 - K) \cdot pot + K \cdot pot_{raw} \tag{4}$$

where is the pot signal processed by the filter, K is the filter coefficient, pot_{raw} is the value on the analog input of the controller. The K coefficient determines which fluctuations are considered too high and require suppression, and which should be skipped with minimal changes. The higher the K, the weaker the filter, and the higher the fluctuations it can skip. Figure 5 shows the original signal (blue) and the signal processed by the low-pass filter (red). This trace was taken from the card reader at rest.

Below is a power trace of a SIM card with random input parameters, which was processed by a low-pass filter [15]. This trace identifies 4 similar parts, each containing 10 rounds. Respectively, the result of the experiment can be argued that this SIM card correlates with the MILENAGE algorithm [6] (Fig. 6).

Fig. 5 Signal tracing at rest

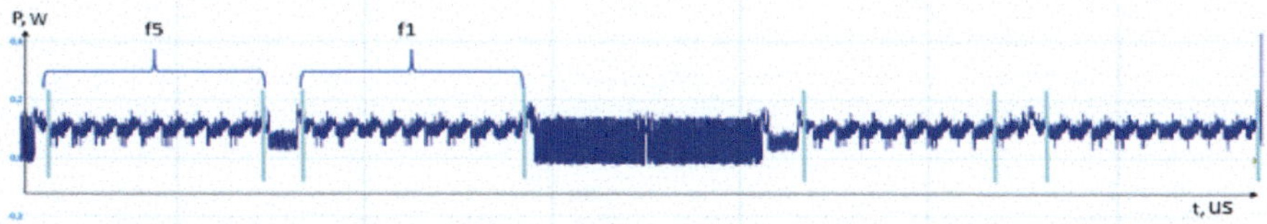

Fig. 6 Power trace

4 Conclusion

Finally, it is important to underline that enhancing the security of SIM cards against SCA is a crucial task given the rapid advancement of digital technologies and the widespread use of mobile communications. With the increasing complexity and frequency of cyberattacks, including those targeting mobile devices, the security of SIM cards becomes a key component of overall cybersecurity. The effectiveness of SIM card protection not only impacts the security of users' personal data but also the integrity of mobile communication networks as a whole.

During the experiment, a hardware and software complex was developed to determine the authentication algorithm used in LTE SIM cards by analyzing energy consumption. The complex was used to identify the SIM card authentication algorithm, indicating that the studied LTE SIM cards use an unsecured version of the MILENAGE algorithm. The data obtained suggests that these SIM cards may be vulnerable to DPA attacks. In future research, we plan to use DPA to extract Ki values from SIM cards to gain a more complete understanding of their security vulnerabilities.

Acknowledgements This research was funded by the Ministry of Science and Higher Education of the Russian Science Foundation № 075-00003-24-02 from 08.02.2024 (Project "Goszaadanie", FSEE-2024-0003).

References

1. Kocher P, Jaffe J, Jun B (Dec 1999) Dierential power analysis. In: Advances in cryptology—CRYPTO' 99, pp 388–397
2. Gierlichs B, Batina L, Tuyls P, Preneel B (2008) Mutual information analysis. In: A generic side-channel distinguisher conference paper CHES 2008, pp 426–442
3. Bertoni G, Daemen J, Debande N, Le T-H, Peeters M, Van Assche G (Dec 2012) Power analysis of hardware implementations protected with secret sharing. In: 2012 45th annual IEEE/ACM international symposium on conference: microarchitecture workshops (MICROW)
4. Kocher P, Jaffe J, Jun B, Rohat P (Mar 2011) Introduction to differential power analysis. J Cryptogr Eng, 5–27
5. ETSI-TS 135 205 universal mobile telecommunications system (UMTS); LTE; 3G security; specification of the MILENAGE algorithm set: an example algorithm set for the 3GPP authentication and key generation functions f1, f1*, f2, f3, f4, f5 and f5*; document 1: general
6. Liu J, Yu Y, Standaert F-X, Guo Z, Gu D, Sun W, Ge Y, Xie X (Sep 2015) Small tweaks do not help: differential power analysis of MILENAGE implementations in 3G/4G USIM cards. In: Conference: European symposium on research in computer security
7. Gustov V, Levina A (2021) Electromagnetic fields as a sign of side-channel attacks in GSM module. In: 2021 11th IFIP international conference on new technologies, mobility and security (NTMS), Paris, France, pp 1–5. https://doi.org/10.1109/NTMS49979.2021.9432678
8. Ometov A, Levina A, Borisenko P, Mostovoy R, Orsino A, Andreev S (2017) Mobile social networking under side-channel attacks: practical security challenges. IEEE Access 5:2591–2601. https://doi.org/10.1109/ACCESS.2017.2665640
9. Levina A, Borisenko P, Mostovoy R (2017) SCA as mobile security threat. In: 2017 20th conference of open innovations association (FRUCT), St. Petersburg, Russia, pp 236–241. https://doi.org/10.23919/FRUCT.2017.8071317
10. ETSI-TS 135 206 universal mobile telecommunications system (UMTS); 3G security; specification of the MILENAGE algorithm set: an example algorithm set for the 3GPP authentication and key generation functions f1, f1*, f2, f3, f4, f5 and f5*; document 2: algorithm specification (3GPP TS 35.206 version 7.0.0 Release 7)
11. Borgaonkar R, Hirschi L, Park S, Shaik A (Jul 2019) New privacy threat on 3G, 4G, and upcoming 5G AKA protocols. CC BY-NC-ND, pp 108–127
12. Daemen J, Rijmen V (2002) The design of Rijndael: AES the advanced encryption standard. Springer
13. ISO/IEC 7816-4 (In press 2014) Identification cards—integrated circuit cards—part 4: organization, security and commands for interchange (IDT), pp 52–60
14. ISO/IEC 14443-4 (2015) Identification cards—contactless integrated circuit cards—proximity cards—part 4: transmission protocol (IDT)
15. Schneider T, Moradi A (Jan 2015) Leakage assessment methodology. In: CHES 2015, pp 495–513

Enhanced Telecom Customer Churn Prediction Through Genetic Algorithm-Based Feature Selection and Machine Learning Techniques

Zou Yuxin, Ghulam Mohi-ud-din, Deng Zhiwei, Xiong Yang, Xia Ling Lin, Chen Ai, and Hu Min

Abstract In the telecommunications industry, a large customer base generates huge amounts of data every day. In this case, acquiring new customers is more expensive than keeping existing ones, and attrition refers to the process of transferring customers from one company to another within a given prescribed time period. With the increasing number of telecom operators, the forecast of telecom customer churn has become an important demand. However, due to the large amount of data, sparse and uneven, telecom customer churn forecasting has always been a complex task. In this paper, Genetic Algorithm (GA) is used for feature selection, and Random Forest (RF), Decision Tree (DT) and other classifiers are used for classification. The experiment was validated on Orange Telecom's customer churn dataset. And the experimental results show that the RF model based on GA feature selection has the best processing effect on customer churn data, and the accuracy of this model is 0.9535.

Z. Yuxin · D. Zhiwei · X. Yang
College of International Education, Nanchang Hangkong University, Nanchang, People's Republic of China
e-mail: 3156362032@qq.com

D. Zhiwei
e-mail: 598702819@qq.com

X. Yang
e-mail: 2829727259@qq.com

G. Mohi-ud-din (✉) · X. L. Lin · C. Ai
School of Software, Nanchang University, Nanchang, People's Republic of China
e-mail: Mohiuddin@ncu.edu.cn

X. L. Lin
e-mail: xialinglin@ncu.edu.cn

C. Ai
e-mail: lisachen@ncu.edu.cn

H. Min
Network and Information Technology Center, Jiangxi University of Traditional Chinese Medicine, Nanchang, People's Republic of China
e-mail: humin@jxutcm.edu.cn

© The Author(s), under exclusive license to Springer Nature Switzerland AG 2025
S. Femmam and P. Lorenz (eds.), *Recent Advances in Communication Networks and Embedded Systems*, Lecture Notes on Data Engineering and Communications Technologies 256, https://doi.org/10.1007/978-3-031-92449-1_12

Keywords Customer churn · Feature selection · Genetic algorithm · Random forest · Machine learning

1 Introduction

Churn, in which customers switch to other products or services, is becoming more common as product and service differences shrink and the risk of brand change decreases. Through the study of customer churn, enterprises can maintain a stable customer base, enhance market competitiveness, and improve long-term economic benefits. Therefore, predicting and preventing customer churn has become a key concern for telecom operators.

Our paper presents a feature selection method based on GA and multiple classifiers to predict customer churn accurately. GA, an optimization algorithm simulating natural evolution, effectively selects valuable features. We introduce a dynamic attention weight adjustment mechanism within the GA, enhancing the selection process by adjusting the importance of individuals based on performance. Using the GA-selected features, we employ various classifiers, including RF and DT, for prediction. Experimental results show that our method outperforms traditional methods in processing customer churn data.

All in all, the research in this paper not only provides a new solution for customer churn prediction in telecom industry, but also provides a reference for similar problems in other industries. Through effective feature selection and classifier combination, enterprises can better understand customer behavior, develop more effective customer retention strategies, and ultimately achieve their long-term growth and profitability goals.

2 Related Work

The paper [1] proposed a RF model based on Scatter Search, which was experimentally verified on telecom customer churn dataset and achieved good classification performance. In this paper, the proposed Scatter Search RF model is compared with other classical classifiers such as Naive Bayes, Support Vector Machine (SVM), DT and multi-layer perceptron. The results show that the Scatter Search RF model has the best performance on various performance metrics, reaching an accuracy of 91.90%.

The study [2] discusses the use of a combination of Convolutional Neural Networks (CNNS) and variational autoencoders to predict the churn of telecommunications customers. The study aims to address the challenge of accurately predicting customer churn in the telecommunications industry, which is critical to maintaining a competitive edge. Hemlata Jain et al. adopted a CNN-based model with a variational autoencoders component to capture underlying patterns and characteristics in telecommunications data, thereby improving the performance of customer churn prediction. Key findings from the study show that the proposed approach outperforms traditional machine learning models in terms of accuracy, precision, recall and F1-score, making it a promising solution for telcos to better understand and manage customer churn.

The study [3] proposed a customer churn prediction model based on machine learning, which was applied to the telecom industry. The authors used the IBM customer loyalty dataset on Kaggle and compared the performance of three models: logistic regression, gradient lift, and RF. The results show that the best model can predict whether a prospect will leave the company with a 93.00% probability. These models can be applied to develop customer relationship management policies and overall sales strategies.

The research [4] proposes a hybrid method combining cluster analysis and DT to predict and analyze customer churn in the telecommunications industry. First, the k-means clustering algorithm is used to identify the market segments with the highest customer churn rate. The chi-squared automatic interaction detector DT is then applied to analyze the customer characteristics of these high-churn segments and extract rules that can be used to develop customer retention strategies. The advantage of this hybrid approach is that it can develop targeted marketing strategies for different vulnerable customer groups. Compared with the model using company-wide customer data, this method can better reflect the churn characteristics of different customer groups. In addition, the research proposes a continuous customer churn management process, suggesting that companies incorporate market segmentation into customer relationship management and continuously optimize according to the effectiveness of the rules. At the same time, providing personalized retention incentives for high churn risk customers is also an important recommendation. Overall, the study suggests an innovative hybrid approach that could help telcos better identify and manage churn prone customers, thereby improving customer retention.

The paper [5] proposes a prediction model of telecom customer churn based on semi-supervised learning. The model first trains the initial model on a small amount of labeled data using supervised learning algorithms such as RF, SVM, logistic regression, etc., Then the pseudo-labeling technology is used to incorporate the unlabeled data into the training to further optimize the model performance. The experimental results show that this semi-supervised learning method can significantly improve the accuracy of customer churn prediction, especially on gradient lift and extreme gradient lift classifiers.

The study [6] proposes a new customer churn prediction system based on ensemble learning and feature grouping. The system uses a stacked model of four machine learning algorithms—XGBoost, logistic regression, DT, and Naive Bayes—to improve prediction accuracy. At the same time, by equidistant grouping of customer behavior and consumption characteristics to construct new features, the potential information is discovered and the model performance is further improved.

The paper [7] introduces the use of business analytics classification model for intelligent customer churn prediction in the telecommunications industry. The main goal is to predict the overall churn rate, accurately identify churn customers, and provide a course of action to prevent business loss. The authors used four well-known predictive models: RF, DT, logistic regression, and k-Nearest Neighbor (kNN), applied to a dataset of Orange Company customers from Kaggle. The results showed that RF outperformed other models in terms of accuracy and F1-score, making it a clear winner.

In this paper [8], a variety of machine learning algorithms, including DT, RF, kNN, Naive Bayes and XGBoost, are used to predict customer churn. Among them, RF and XGBoost algorithms perform best in terms of accuracy and time complexity. In addition, hyperparameter optimization techniques such as grid search are used to improve the model performance, and evaluation indexes such as precision rate-recall curve and F1-score are used to evaluate the model comprehensively.

At present, many researches have explored the methods of customer churn prediction, however, there are still some shortcomings. Traditional feature selection methods may not adequately capture key features in customer churn data, resulting in insufficient accuracy of predictive models. In addition, a single feature selection approach may not be optimal when dealing with complex customer churn issues. To address these shortcomings, we propose an innovative GA feature selection method, combined with a "dynamic attention weight adjustment" mechanism, and apply it to multiple classifiers to more accurately predict customer churn.

3 Proposed Method

In our study, we utilized a GA-based feature selection method applied across different classifiers to analyze customer churn data. Initially, we removed highly correlated and unnecessary columns from the dataset and performed label encoding and one-hot encoding for categorical features, while standardizing numerical features. GA was then employed for feature selection, optimizing the selection process by simulating the biological evolution operations of selection, crossover, and mutation. The goal was to identify the best combination of features to enhance each classifier's performance. Each individual in the GA population was represented by a gene sequence, where each gene bit indicated whether a corresponding feature was selected. The GA dynamically adjusted attention weights based on individual performance to improve selection efficiency. The best-selected features were then used to train classifiers, with hyperparameters optimized via grid search. Model performance on the test set was evaluated using metrics such as accuracy, precision, recall, and F1-score. The results demonstrated that this method effectively improved model classification performance. The model is shown in Fig. 1.

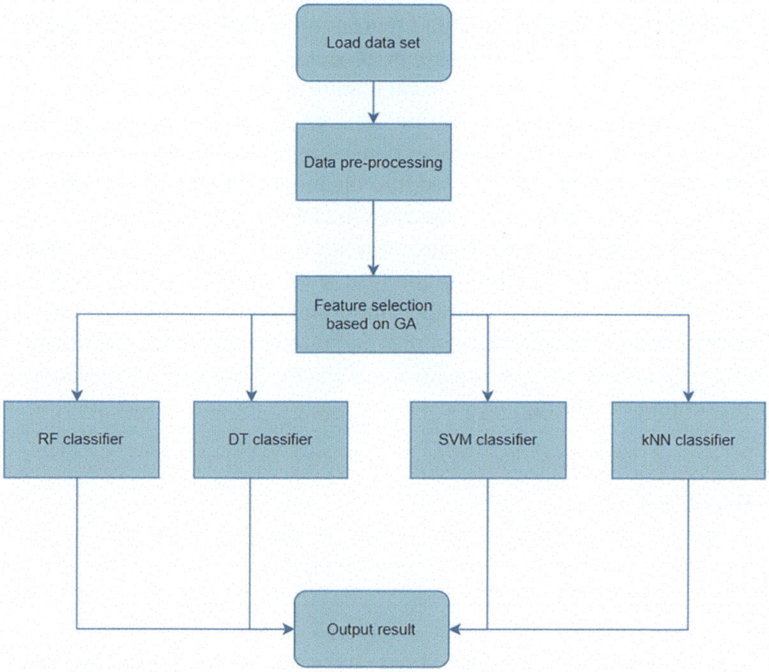

Fig. 1 Proposed methodology

3.1 *Orange Telecom's Churn Dataset*

Orange Telecom's Churn Dataset is a classic dataset commonly used in data science and machine learning to study and predict customer churn. The dataset is derived from real telecom company customer data and contains rich information on customer behavior and service usage. By using Orange Telecom's Churn Dataset for research, we can not only improve our understanding of customer attrition issues, but also apply a variety of data science and machine learning techniques to solve real-world problems, thereby bringing immediate business value to businesses. These advantages make this data integration ideal for studying customer churn and data science teaching.

3.2 *Classifier*

3.2.1 Random Forest

RF [9] is an ensemble learning method proposed by Leo Breiman and Adele Cutler in 2001 to solve classification and regression problems. RF is a supervised learning algorithm suitable for classification and regression problems. It can be used in data mining, computer vision, natural language processing and other fields. RF are built on the basis of decision trees. An important feature of RF is that it can reduce the overfitting of decision trees due to overfitting data, thereby improving the performance of models. A RF is an integrated model consisting of many decision trees. Its core idea is that when the training data is fed into the model, instead of building one large decision tree with the entire training data set, the RF builds multiple smaller decision trees with different subsets and feature attributes, and then combines them into a more powerful model. By combining the results of multiple decision trees, RF can enhance the effect of the model. Another important feature of a RF is that each subset is built from randomly selected samples and randomly selected feature attributes. This randomization reduces the sensitivity of the decision tree to the training data, thereby preventing overfitting.

3.2.2 Decision Tree

DT [10] is a basic classification and regression method. The DT model is a tree-like structure, which represents the process of classifying instances based on features in classification problems. It can be considered as a set of if-then rules, or as a conditional probability distribution defined on the feature space and class space. The main advantage is that the model is readable and the classification speed is fast. During learning, the DT model is built according to the principle of minimization of loss function by using training data. When forecasting, the new data is classified by DT model.

3.2.3 Support Vector Machine

SVM [11] is a powerful and flexible supervised learning algorithm [12] for classification [13], regression, and anomaly detection tasks. The main idea is to find an optimal hyperplane that separates data points of different classes and maximizes the spacing of classification boundaries. The SVM model is show in Fig. 2.

3.2.4 K-Nearest Neighbor

kNN [14] method was first proposed by Cover and Hart in 1968, is a relatively mature method in theory and one of the simplest machine learning algorithms. k nearest neighbor means that each sample can be represented by its nearest k neighbor values. kNN is a commonly used supervised learning method, and its working mechanism is very simple: Given a test sample, k training samples in the training set close to its quantity can be found based on some distance measure, and then prediction can be made based on the information of these k "neighbors".

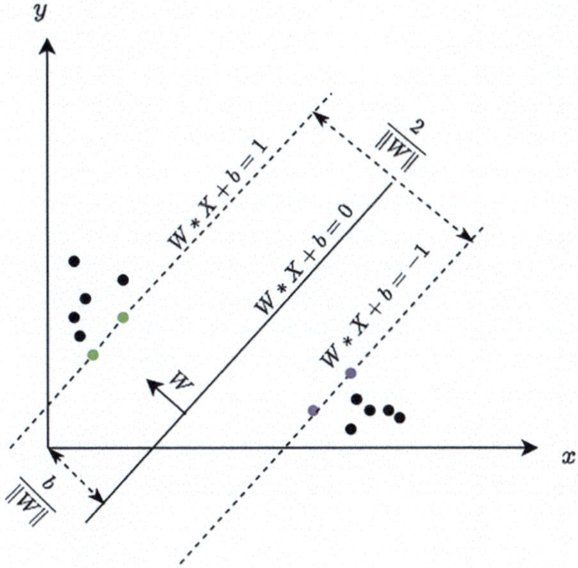

Fig. 2 Working of SVM diagram

3.3 *Dimension Reduction*

3.3.1 Genetic Algorithm

GA [15] is a heuristic global optimization technique, which aims to simulate the process of biological evolution in nature to solve complex optimization problems. This method was first proposed by American computer scientist John Holland in the late 1960s. Based on the principles of natural selection and genetics, GA uses a global adaptive probabilistic search mechanism to find the optimal solution of a problem. GA start with a randomly generated initial population consisting of multiple individuals, each of which is a coded representation of the problem solution, often called a chromosome. The chromosome contains the characteristics of the solution to the problem, and its internal expression (genotype) consists of a set of genes that determine the external performance of the individual (phenotype). In the initialization phase of the algorithm, the mapping process from phenotype to genotype must be completed, that is, the encoding of the problem solution to the chromosome. After initializing the population, the GA evolves through generations to gradually improve the solution by simulating the natural evolution process. In each generation, the fitness is evaluated for each individual in the population, and the design of the fitness function depends on the specific problem domain and reflects the quality of the solution. Individuals with high fitness have a greater probability of being selected for reproduction, which is achieved by selection operators, and commonly used selection strategies include roulette wheel selection and tournament selection, among others. The selected individuals undergo genetic operations, including crossover and mutation. The crossover operation generates new individuals by combining partial genes of two or more individuals, simulating paired reproduction in nature. The mutation operation introduces new genetic information by randomly changing the genes in the chromosome, which helps to maintain the diversity of the population and prevent the algorithm from falling into the local optimal solution prematurely. After the generation of new individuals, to keep the population size constant, the replacement strategy is usually adopted, and the individuals with lower fitness are replaced, while the elite retention strategy ensures that some individuals with the highest fitness are retained. This process is repeated, the population evolves gradually, and the individuals gradually tend to better solutions. Finally, the algorithm performs iterations until a predefined termination criterion is satisfied, such as reaching a certain fitness threshold or an upper bound on the number of iterations. And the best individuals selected from the final population are decoded to provide a near-optimal solution to the problem.

3.3.2 Principal Component Analysis

PCA [16] is a commonly used unsupervised learning algorithm [17] for data reduction and feature extraction. It was first proposed by Karl Pearson [18] in 1901, and later several scholars independently proposed it. The main goal of PCA is to project the original high-dimensional data into a low-dimensional space through linear transformations while preserving the information of the original data to the greatest extent possible.

3.4 Experimental Setup

This experiment was conducted using a personal computer equipped with the following configurations: Operating system: Windows 11. Processor: AMD Ryzen 7 5800H with Radeon Graphics 3.20 GHz. RAM: 16.0GB (15.4GB available). System type: 64-bit operating system, x64 based processor.

4 Results and Discussion

In this study, we used Orange Telecom's Churn Dataset to evaluate the performance of four classifiers (RF, DT, SVM, and kNN) under both GA and PCA feature selection methods. Performance metrics include accuracy, precision, recall and F1-score.

For RF, feature selection using GA is superior to PCA on all metrics. The accuracy of GA was 0.9535, while that of PCA was 0.8936. Accuracy increased from 0.8000 for PCA to 0.9571 for GA, recall and F1-score also increased from 0.3368 and 0.4741 to 0.7053 and 0.8121, respectively.

Table 1 Performance metrics with GA-based dimensionality reduction

Metric	RF	DT	SVM	kNN
Accuracy	0.9535	0.9475	0.9415	0.9145
Precision	0.9571	0.8571	0.8889	0.7969
Recall	0.7053	0.7579	0.6737	0.5368
F1-score	0.8121	0.8045	0.7665	0.6415

Table 2 Performance metrics with PCA-based dimensionality reduction

Metric	RF	DT	SVM	kNN
Accuracy	0.8936	0.8786	0.9160	0.8951
Precision	0.8000	0.6667	0.7746	0.9032
Recall	0.3368	0.2947	0.5789	0.2947
F1-score	0.4741	0.4088	0.6627	0.4444

DT classifiers showed similar trends, with GA accuracy increasing from 0.8786 to 0.9475, accuracy from 0.6667 to 0.8571, recall from 0.2947 to 0.7579, and F1-score significantly improving from 0.4088 to 0.8045. This indicates that GA can improve the predictive ability of DT by selecting more informative features.

For SVM, the difference between GA and PCA was less pronounced, but GA performed better overall. GA's accuracy was 0.9415, while PCA's was 0.9160, increasing accuracy from 0.7746 to 0.8889, recall and F1-score from 0.5789 and 0.6627 to 0.6737 and 0.7665, respectively.

For kNN classifier, GA has a significant improvement compared with PCA. GA accuracy improved from 0.8951 to 0.9145, but accuracy decreased from 0.9032 to 0.7969, while recall and F1-score improved from 0.2947 to 0.5368 and 0.6415, respectively. This shows that GA improves the ability of kNN classifier to correctly identify positive examples.

In general, GA is generally superior to PCA in feature selection, especially in improving recall rate and F1 score. These results show that GA can more effectively select features that help improve the performance of the classifier. The results are show in Tables 1 and 2.

4.1 GA Versus PCA

In our study, we compare the feature dimensionality reduction performance of GA and PCA on different classifiers. It can be seen from the experimental results that GA is superior to PCA in many performance metrics. Especially in DT and SVM classifiers, the advantage of GA is more obvious. This is mainly because GA can more

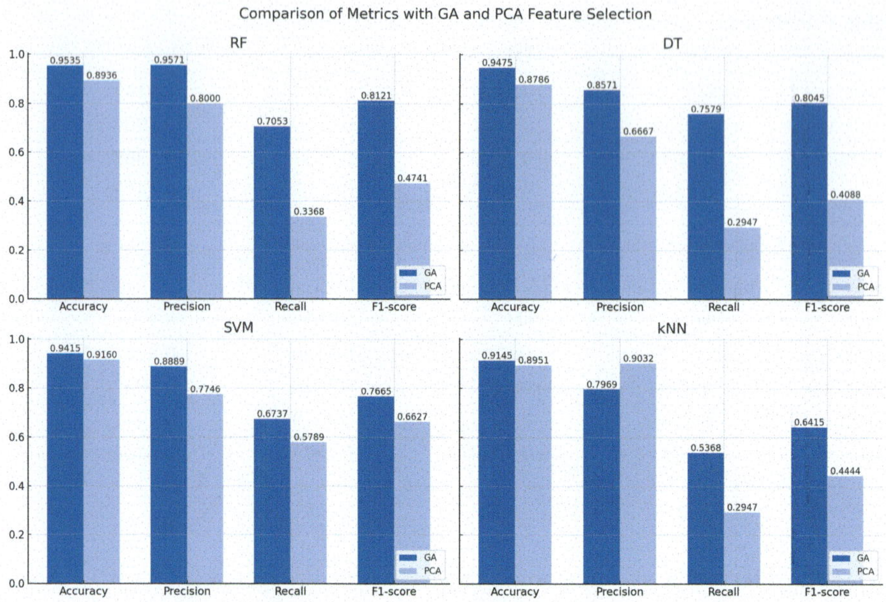

Fig. 3 Comparison of Metrics with GA and PCA Feature Selection

effectively select the features that have significant contributions to the classification, and continuously optimize the feature subset through the evolutionary process, thus improving the overall performance of the model. In contrast, PCA, while able to improve model performance to some extent, is essentially a linear transformation method designed to reduce dimension by maximizing variance. However, this approach may lose some nonlinear feature information that is critical to the classification task. By simulating natural selection and genetic variation, GA is able to retain and optimize these important nonlinear features to better capture important information in the data. These results show that GA, as a feature selection method, can mine the key features in the data more comprehensively and significantly improve the performance of the classifier, so it is more suitable for processing complex classification tasks. The comparison of metrics with GA and PCA feature selection is show in Fig. 3.

5 Comparision

Varun et al. [1] used Orange Telecom's Churn Dataset, applied correlation-based feature selection and scatter search techniques, and then used RF for classification. The optimal RF model accuracy of this method is 91.90%. In contrast, our approach combines GA for feature selection and uses multiple classifiers for classification and our experimental results show that the accuracy of the RF classifier using GA for feature selection is 95.35%, which is significantly better than their method.

Jain et al. [2] used a combination of CNNS and variational autoencoders to predict customer churn. The method achieved 88.00% accuracy. In contrast, our method achieves 95.35% accuracy on RF classifiers using GA for feature selection, which is significantly better than the method they proposed, indicating that our feature selection method and classifier perform better in customer churn prediction.

The study [3] proposes a machine learning-based customer churn prediction model, which is applied to telecom companies. The authors used the IBM customer loyalty dataset on Kaggle and compared the performance of three models: logistic regression, gradient lift, and RF. The results show that the Gradient Boosting model is the best model, and its prediction accuracy can reach 94.5%. In contrast, our experimental results show that the accuracy of GA combined with RF reaches 95.35%, which is not only higher than the RF model in their study (93.4%), but also higher than the best Gradient Boosting model in their study (94.5%), which shows the superiority of our method.

Gohil et al. [7] adopted logistic regression, RF, DT and kNN classification models to predict customer churn. The RF model in their study has an accuracy of 0.951. In contrast, our experimental results show that the accuracy of GA combined with RF reaches 0.9535, which is higher than the method in their study, indicating that our method has certain advantages in accuracy.

Gowd et al. [8] used a variety of machine learning algorithms such as RF, DT, kNN and Naive Bayes to predict customer churn. Their method achieved an RF accuracy of 95.20% by using Grid Search for hyperparameter optimization. In contrast, our approach uses GA for feature selection methods and combines multiple classifiers (RF, DT, SVM, and kNN) for customer churn prediction. In the case of feature selection using GA, the accuracy of the RF classifier reached 95.35%, which is higher than the 95.20% accuracy of the RF model in their study. In addition, our study also showed that the accuracy of GA combined with DT, SVM and kNN was 94.75%, 94.15% and 91.45%, respectively. These results show that the application of GA in feature selection significantly improves the performance of classifiers, especially for RF classifiers. Our approach not only shows higher accuracy on RF classifiers, but also shows broader applicability and greater stability by using multiple classifiers. Detailed data are shown in Table 3.

6 Conclusion and Future Work

Our study develops a customer churn prediction model based on machine learning for telecom industry. Our method uses GA for feature selection, combines with a variety of classifiers for prediction, and compares the performance of two dimensionality reduction methods, GA and PCA. The experimental results show that the accuracy of the RF classifier using GA for feature selection reaches 95.35%, while the accuracy of the RF using PCA for feature selection is only 89.36%. This shows

Table 3 Comparison with other works

Work	Model	Accuracy (%)
Varun et al. [1]	SVM	70.94
	DT	62.44
	RF	91.90
Jain et al. [2]	CNN	88.00
Todevski et al. [3]	RF	93.40
	GB	94.50
Gohil et al. [7]	kNN	87.10
	DT	93.40
	RF	95.16
Gowd et al. [8]	DT	93.40
	RF	95.20
	kNN	88.30

that GA is significantly better than PCA in feature selection. In addition, other classifiers such as DT, SVM and kNN also perform better than PCA under GA feature selection. These results demonstrate the superiority of GA in customer churn prediction and demonstrate its potential in improving classifier performance. Despite the remarkable results achieved in this study, there are still some shortcomings: First, the computational complexity of GA is high, and the feature selection process consumes a long running time, which may become a bottleneck on large-scale data sets. Secondly, this study mainly uses static data sets, and the performance of the model in processing real-time streaming data has not been verified.

Future studies can further optimize the feature selection process and explore more feature selection algorithms to improve the predictive performance of the model. In addition, more diverse data sets can be considered to verify the generality and robustness of the model. At the same time, the combination of deep learning techniques, such as CNNS and recurrent neural networks, may lead to higher prediction accuracy and greater model generalization ability.

References

1. Varun A, Babu T, Nair RR, Kishore S (2023) Telecom churn classification using scatter search and random forest classifier. In: Proceedings of the 2023 international conference on evolutionary algorithms and soft computing techniques (EASCT), pp 1–6
2. Jain H, Khunteta A, Srivastava S (2022) Telecom churn prediction using CNN with variational autoencoder. Smart Syst: Innov Comput: Proc SSIC 2021:583–600
3. Todevski D, Svrtinov VG (2021) Machine learning model for customer churn. KNOWLEDGE-Int J 47(5):887–892
4. Pejić Bach M, Pivar J, Jaković B (2021) Churn management in telecommunications: hybrid approach using cluster analysis and decision trees. J Risk Financ Manag 14(11):544

5. Rani B, Kant S (2020) Semi-supervised learning approach to improve machine learning algorithms for churn analysis in telecommunication. Int J Comput Inf Syst Ind Manag Appl 12:11–11
6. Xu T, Ma Y, Kim K (2021) Telecom churn prediction system based on ensemble learning using feature grouping. Appl Sci 11(11):4742
7. Gohil S, Ameria M, Sarobin MVR (2023) An intelligent prediction of customer churn in telecom industry using business analytics classification models. In: Proceedings of the 2023 12th international conference on advanced computing (ICoAC) pp 1–7
8. Gowd S, Mohite A, Chakravarty D, Nalbalwar S (2023) Customer churn analysis and prediction in telecommunication sector implementing different machine learning techniques. In: Proceedings of the first international conference on advances in computer vision and artificial intelligence technologies (ACVAIT 2022) pp 686–700
9. Breiman L (2001) Random forests. Mach Learn 45:5–32
10. Hunt EB, Marin J, Stone PJ (1966) Experiments in induction. Academic Press
11. Yu H, Kim S (2012) SVM tutorial-classification, regression and ranking. Handb Nat Comput 1:479–506
12. Nasteski V (2017) An overview of the supervised machine learning methods. Horizons. b 4(51–62):56
13. Colas F, Brazdil P (2006) Comparison of SVM and some older classification algorithms in text classification tasks. In: IFIP international conference on artificial intelligence in theory and practice. Springer, pp 169–178
14. Cover T, Hart P (1967) Nearest neighbor pattern classification. IEEE Trans Inf Theory 13(1):21–27
15. Lambora A, Gupta K, Chopra K (2019) Genetic algorithm-a literature review. In: Proceedings of the 2019 international conference on machine learning, big data, cloud and parallel computing (COMITCon), pp 380–384. https://doi.org/10.1109/COMITCon.2019.8862255
16. Maćkiewicz A, Ratajczak W (1993) Principal components analysis (PCA). Comput Geosci 19(3):303–342
17. Barlow HB (1989) Unsupervised learning. Neural Comput 1(3):295–311
18. Pearson KF (2019) On lines and planes of closest fit to systems of points in space. Lond Edinb Dublin Philos Mag J Sci Ser 6, 53

The manufacturer's authorised representative in the EU is Springer Nature Customer Service Centre GmbH, Europaplatz 3, 69115 Heidelberg, Germany. If you have any concerns regarding our products, please contact ProductSafety@springernature.com

Printed and bound by CPI Group (UK) Ltd, Croydon, CR0 4YY
05/10/2026
02228158-0001